Mathematical Theories of Distributed Sensor Networks

Sitharama S. Iyengar · Kianoosh G. Boroojeni
N. Balakrishnan

Mathematical Theories of Distributed Sensor Networks

Sitharama S. Iyengar
Kianoosh G. Boroojeni
Florida International University
Miami, FL
USA

N. Balakrishnan
Indian Institute of Science
Bangalore
India

ISBN 978-1-4419-8419-7 ISBN 978-1-4419-8420-3 (eBook)
DOI 10.1007/978-1-4419-8420-3
Springer New York Heidelberg Dordrecht London

Library of Congress Control Number: 2014936867

Printed on acid-free paper

Springer is part of Springer Science+Business Media (www.springer.com)

Preface

Distributed Sensor Networks (DSN) are large-scale, autonomous, resource-constrained systems used for gathering data in an intelligent manner. Described by Business week as one of the twenty-first century, DSN technology is stepping out its cradle of laboratories and research papers, and finding increasing number of applications.

Often powered by advances in the miniaturization of Microelectronic and Mechanical Structure (MEMS), a DSN is composed of many tiny sensor nodes, all of which have capabilities for sensing, computing, and communication. Sensor nodes are often low cost, low power, and small size, and frequently remain untethered and unattended after deployment. Wireless Sensor Network (WSN) typically consists of large number of various kinds of unstructured environments. Usually, a sensor node has a communication range of less than 100 ft. Sensor nodes can be deployed on the ground, in the soil, underwater, in the air, in vehicles, or in buildings.

The detection and identification methods must be robust and secure for persistent and pervasive operations under uncertainty, resource constraints, and known and unknown operational, environmental, and adversarial perturbations. This necessitates the development of sensor networks that autonomously form collaborative clusters for reliable time critical response to natural and man-made disasters and to more general battlefield environments and are backed by strong processing power of cyber systems.

In this monograph, the *Mathematical Theories of Distributed Sensor Networks* are presented. In Chap. 1, we provide an introduction to the DNS. This Introduction chapter has been designed to clarify the importance and position of the area of sensor networks in the context of the network design as its containing field of research.

Energy efficiency is one of the most critical issues in DNS. Efficient time complexity of algorithms which are running in different parts of a DSN for various purposes has a vital role in optimizing the energy consumption of the network. For example, a poor-quality routing algorithm may lead to node congestion and a huge energy wastage. In the "Optimization Methods" part, we will address two famous optimization problems in the context of DSNs in detail: art-gallery problem and wideband source localization using acoustic sensor networks.

One of the most famous *coverage problems* is called the *gallery-guarding* problem or generally, *Art-Gallery Problem.* In Chap. 2, we will see an efficient approximation algorithm which makes an acceptable solution for the gallery-guarding problem in 3-D when the boundary of guarded region is in the form of a polygonal mesh.

Another optimization problem which will be addressed in this book is the wideband source localization using acoustic sensor networks in the presence of nonuniform noise variances.

In Chap. 3, we present a solution to the problem based on two source localization algorithms called Stepwise-Concentrated Maximum-Likelihood (SC-ML) and Approximately Concentrated Maximum-Likelihood (AC-ML).

In Part II of the monograph, we focus on the coverage and connectivity problems in the context of DNS. Coverage problem is one of the fundamental issues in DSNs. In this problem, the goal is to determine how well a set of sensors can monitor a given area. In addition, the connectivity problem in DNS is to determine whether the graph representation of the network is connected or not.

In Chap. 4, we address the coordinate-free coverage problems in DSNs via Homology.[1] Moreover, in Chap. 5, some discussions regarding the coverage assessment and target tracking in 3-D domains will be presented.

Finally, in Part III (last) of the book, an interesting security problem in DSNs will be addressed. More specifically, we propose a novel stochastic preserving scheme of location privacy either for a static or for a mobile sensor node. After describing the proposed scheme, we present the privacy assessment of the method using some mathematical tools.

Features

Here are the unique aspects of our book which address the oblivious network routing problems:

(1) The book specifies the importance and position of the DNS in the context of the network design as it contains the field of research. To do this, numerous concepts in this area are defined precisely using the mathematical tools.
(2) The book provides the basics and mathematical foundations needed to analyze and address many problems in DNS. More specifically, it introduces some advanced data structures and tools (in graph theory and probability theory) which will be deployed for analysis and design of high-performance DNS.

[1] Homology is a certain general procedure to associate a sequence of abelian groups or modules with a given mathematical object such as a topological space or a group.

Intended Audience

This monograph is suitable for senior undergraduate students, graduate students, and the researchers working in the related areas.

Authors' Credentials

S. S. Iyengar is a Distinguished Ryder Professor and Director of the School of Computing and Information Sciences at the Florida International University and is the founding Director of the FIU-Discovery Lab. Iyengar is a pioneer in the field of distributed sensor networks/sensor fusion, computational aspects of robotics, and high-performance computing. Iyengar has published over 500 research papers and has authored/co-authored/edited 18 books published by John Wiley and Sons, Prentice Hall, CRC Press, Springer Verlag, etc. These publications have been used in major universities all over the world.

His research publications are on the design and analysis of efficient algorithms, parallel computing, sensor networks, and robotics. In 2014, Dr. Iyengar was elected to the rank of NAI (National Academy of Inventors) Fellow for demonstrating a highly prolific sprint of innovation in creating/facilitating outstanding inventions that have made a tangible impact on quality of life, economic development, and the welfare of society.

In addition, Dr. Iyengar was awarded the 2014 IBM Faculty Award for fostering collaboration between researchers at leading universities worldwide and those in IBM Research, development and services organizations and prompting courseware and curriculum innovation to stimulate growth in disciplines and geographies that are strategic to IBM. He is also a member of the European Academy of Sciences, a Fellow of IEEE, a Fellow of ACM, a Fellow of AAAS, and a Fellow of Society of Design and Process Program (SPDS), Fellow of Institution of Engineers (FIE), awarded a Distinguished Alumnus Award of the Indian Institute of Science, Bangalore, and was awarded the IEEE Computer Society Technical Achievement for the contributions to sensor fusion algorithms and parallel algorithms. He is a Golden Core member of the IEEE-CS, and he has received a Lifetime Achievement Award conferred by the International Society of Agile Manufacturing (ISAM) in recognition of his illustrious career in teaching, research, and a lifelong contribution to the fields of Engineering and Computer Science at Indian Institute of Technology (BHU). Iyengar and Nulogix were awarded in the 2012 Innovation 2 Industry (i2i) Florida competition. Iyengar received Distinguished Research Award from Xaimen University, China, for his research in Sensor Networks, Computer Vision, and Image Processing. Iyengar's landmark contributions with his research group are the development of grid coverage for surveillance and target location in DNS and Brooks Iyengar fusion algorithm.

He has also been awarded honorary Doctorate of Science and Engineering from an institution. He serves on the advisory board of many corporations and universities in the world. He has served on many National Science Boards such as NIH—National Library of Medicine in Bioinformatics, National Science Foundation review panel, NASA Space Science, Department of Homeland Security, Office of Naval Security, and many others. His contribution was a centerpiece of this pioneering effort to develop image analysis for our science and technology and to the goals of the US Naval Research Laboratory. The impact of his research contributions can be seen in companies/National Labs like Raytheon, Telecordia, Motorola, the United States Navy, DARPA agencies, etc. His contribution includes DARPAS's program demonstration with BBN, Cambridge, Massachusetts, MURI, researchers from PSU/ARL, Duke, University of Wisconsin, UCLA, Cornell University, and LSU.

He is also the Founding Editor of International Journal of Distributed Sensor Networks. He has been the Editor for many IEEE Journals (IEEE Transactions on Computers, etc.). He is presently the Editor of ACM Computing Surveys and other journals. Also he is the founding director of the FIUs Discovery Laboratory. His research work has been cited over extensively in Wikipedia and in other places. Iyengar has graduated over 45 Ph.D. students, 100s of Masters Students, and a large number of postdoctoral fellows at various institutions in the world. He also had many undergraduate students working on his research projects.

His fundamental work has been transitioned into unique technologies. The impact of his work is significant and can be seen in several areas of computer science. His work had been the fundamental in the development of technological innovations in several organizations around the world. All through his three-decade long professional career, Iyengar has devoted and employed algorithmic morphology in a unique way for quantitative understanding of computational processes for many applications.

Kianoosh Gholami Boroojeni is a Ph.D. student of computer science at FIU. He received his B.Sc. in University of Tehran, Iran (2012). His research interests include algorithms and combinatorial optimization.

During the first year of Kianoush's graduate years, he has submitted many peer-reviewed journal papers and a book to the Springer publications. Currently, Kianoosh is collaborating with Dr. S. S. Iyengar on some optimization problems in the context of oblivious network design.

N. Balakrishnan is an Associate Director and Professor at Department of Aerospace Engineering and Supercomputer Education and Research Centre, Indian Institute of Science. His research interests include "numerical electromagnetic," "multi-parameter radars," and "signal processing."

Professor Balakrishnan is the Fellow of Third World Academy of Sciences, Indian National Science Academy, Indian Academy of Sciences, Indian National Academy of Engineering, National Academy of Sciences, Allahabad, and Institution of Electronic and Telecommunication Engineers. His publications include 19 books and many peer-reviewed journal papers.

Acknowledgments

This work has evolved from our research on oblivious network design in distributed environments. It has been made possible, thanks to funding from office of Naval Research, National Science Foundation, and several other resources. The authors would like to thank Ivana Rodriguez for her helpful comments in improving the readability and presentation of this monograph. Professor Iyengar wishes to thank many students including Srivathsan Srinivasagopalan, Vasanth Iyer, Richard Brooks, Costas Busch, and others for their contributions. Kianoosh G. Boroojeni would like to express his sincere gratitude to his family and his colleagues Farhad Hemmati, Shekoufeh Mokhtari, Navid Khazraei, and Maziar Kordmahaleh for their continuous inspiration and support.

Contents

Part III Security Issues

Chapter 1
Introduction to Distributed Sensor Networks

Currently, detection and tracking systems use a large number of different types of sensors. Because of the relatively low cost of sensors, many duplicate sensors of the same type are used to insure increased fault tolerance. The common practice is to assign each sensor of sensor cluster to handle on specific task. For example, while tracking multiple target, one sensor cluster is assigned to track one target only and any information it may collect about other targets is not utilized.

Thus, there is a great deal of interest in integrating all the sensors so that the information can be effectively utilized. The sophisticated demands made on the tracking and surveillance systems have generated a great deal of interest in developing new architectures that allow for the fusion of the information in the different sensors. For instance, it should be possible to combine the information given by infrared sensors with that given by microwave radars. An such integration of sensors implies the availability of communication networks that allow for the transfer of information between sensors.

More specifically, the design of spatially distributed target detection and tracking systems involves the integration of solutions obtained by solving subproblems in data association and fusion, hypothesis testing, effective computational strategies, etc. We envision a cooperative resolution of the overall problem using the solutions of the subproblems available at local sensors. No single sensor or sensors are really not practicable. Both data collection and control have to be logically and geographically distributed necessitating the sharing of information and the use of cooperative problem solving approaches.

Some portion of this chapter has been reprinted with permission from "Distributed Sensor Networks-Introduction to the Special Section," Iyengar et al. [1]

S. S. Iyengar et al., *Mathematical Theories of Distributed Sensor Networks*,
DOI: 10.1007/978-1-4419-8420-3_1,

1.1 What is a Distributed Sensor Network?

A distributed sensor network (DSN) can be defined as a *set of spatially scattered intelligent sensors designed to obtain measurements from the environment, to abstract relevant information from data gathered, and to derive appropriate inferences from the information gained.* DSNs depend on multiple processors to simultaneously gather and process information from many sources. Interest in theses systems stems from a realization of the limitations imposed by relying on a single source of information to make decisions.

Currently, there has been an increasing interest in the development of DSNs for the process of information gathering. Availability of new technology makes these networks economically feasible. The increased complexity of today's information gathering tasks has created a demand for such networks. These tasks are usually time-critical and rely on the reliable delivery of accurate information. Thus, the search for efficient, fault-tolerant architectures for DSNs has become an important research area in computer science.

1.1.1 Requirements of Distributed Sensor Networks

A DSN is basically a system of connected, cooperating generally diverse sensors that are spatially dispersed. The *major task* of a DSN is to process data possibly noise corrupted, acquired by the various sensors and to integrate it, reduce the uncertainty in it, and produce abstract interpretations of it. Three important facts emerge from such a framework:

(1) The network must have intelligence at each node,
(2) It must accommodate diverse sensors, and
(3) Its performance must not degrade because of spatial distribution.

DSNs are assumed to function under the following conditions:

- Each sensor in the ensemble can see some but not all of the low-level activities performed by the sensor network as a whole.
- Data is perishable, in the sense that information value depends critically upon the time required to acquire and process it.
- There should be limited communication among the sensor processors, so that a communication–computation trade-off can be made.
- There should be sufficient information in the system to overcome certain adverse conditions (e.g., node and link failures) and still arrive at a solution in its specific problem domain.

The successful integration of multiple, diverse sensors into a useful sensor network requires the following:

- The development of methods to abstractly represent information gain from sensors so that this information may easily be integrated.

- The development of methods to deal with possible differences in point of view on frames of reference between multiple sensors.
- The development of methods to model sensor signals so that the degree of uncertainty is reduced.

1.1.2 Communication in DSNs

In a typical DSN, each node needs to fuse the local information with the data collected by the other nodes so that an updated assessment is obtained. Current research involves fusion based on a multiple hypothesis approach. Maintaining consistency and eliminating redundancy are two important considerations. The problem of determining what should be communicated is more important than how this communication is to be effected. An analysis of this problem yields the following classes of information as likely candidates for being communicated: information about the DSN, information about the state of the world, hypothesis, conjectures, and special requests for specific actions. It is easy to see that different classes of information warrant different degrees of reliability and urgency. Further details regarding information fusion in DSNs may be found in [2–8] (adopted from Iyengar et al. [1]).

1.2 Algorithms and Complexity Issues

Energy efficiency is one of the most critical issues in DSNs. Efficient time complexity of algorithms which are running in different parts of a DSN for various purposes has a vital role in optimizing the energy consumption of the network. For example, a poor-quality routing algorithm may leads to node congestion and a huge energy wastage. In Chaps. 2 and 3, we will address two famous optimization problems in the context of DSNs in detail. Here, we introduce these two optimization problems:

1.2.1 Gallery-Guarding Problem in 3-D

One of the common types of *coverage problems* (which will be introduced in Sect. 1.4) is called the gallery-guarding problem or generally, *Art-Gallery Problem* [9]. The main objective of this problem in 3-D regions is to find a set of points like $\mathcal{S}$ in the geometric region $\mathcal{A} \subseteq \mathbb{R}^3$ such that[1]:

$$\forall X \in \mathcal{A}{:}\exists Y \in \mathcal{S}{:}\overline{XY}^3 \subseteq \mathcal{A} \tag{1.1}$$

[1] For every pair of points $X, Y \in \mathbb{R}^3$, $\overline{XY}$ denotes the line segment connecting X and Y.

In other words, any point in region $\mathcal{A}$ can be *visible* from a point in the *solution* set $\mathcal{S}$. It is easy to prove that the following proposition is equivalent to Proposition 1.1:

$$\forall X \in \partial\mathcal{A}{:}\exists Y \in \mathcal{S}{:}\overline{XY} \subseteq \mathcal{A}$$

where $\partial\mathcal{A} \subseteq \mathcal{A}$ denotes the boundary of region $\mathcal{A}$.

In Chap. 2, we will see an efficient approximation algorithm which makes an acceptable solution for the gallery-guarding problem in 3-D when the boundary of region $\mathcal{A}$ is in the form of a polygonal mesh.

1.2.2 Wideband Acoustic Source Localization in the Presence of Nonuniform Noise Variances

Another optimization problem which will be addressed in this book is the wideband source localization using acoustic sensor networks in the presence of nonuniform noise variances.

In Chap. 3, we present a solution to the problem based on two source localization algorithms called stepwise-concentrated maximum likelihood (SC-ML) and approximately-concentrated maximum likelihood (AC-ML).

1.3 Layered Network Architecture

DSNs (especially, the wireless ones) have typically low-power nodes, with limited CPU and memory, and the application requirements and network architecture of DSNs differ greatly from that of traditional computer networks.

In this section, we focus on the network architecture of DSNs using the well-known layered approach. In the layered network architecture, every layer uses the services of lower layers and provides with specific services which will be used by the upper ones. The major differences of DSNs and traditional computer networks will be addressed in each architecture layer.

1.3.1 Physical Layer

Physical layer is the lowest layer in the layered architecture of a network. All of the other four layers rely on the services which this layer provides. The main task of the physical layer is to convert the data (which is intended to be sent through the network) into the electromagnetic waves at any connection source and do the reverse conversion at the connection target. These conversions are called as "modulation" and "demodulation," respectively. In order to propagate data through the network via different media such as air, copper wires, and optical fibers, we need to do modulation/demodulation at the sensor nodes of a DSN to transmit the electromagnetic waves in the connection sessions. In some of the network architectures, data are encrypted before modulation at the connection source and decrypted on the other side of the connection when demodulation occurs.

Regarding the aforementioned power restrictions of DSNs, power optimization is one of the most critical concerns in the physical layer of a DSN. Especially, in wireless sensor networks in which sensors may be connected to limited power resources (like batteries), the physical layer of sensors must be designed to run a low duty cycle, such that the battery is discharged in pulses. In fact, a battery undergoing pulsed discharge generally has less power consumption (in long term) than the one which is constantly being discharged, as the pulsed approach tends to have a form of *charge recovery effect*. Further treatment on this layer can be found in [10–12].

1.3.2 Data Link Layer

In the second network layer, the connecting medium is managed as a resource shared with all the sensor nodes. If the network links are simultaneously used by different connections in a DSN, they may fail to work properly as the collision occurs between different connection waves. This urges a media access control (MAC) protocol to handle different connections which need to use the shared resource at the same time. There are a number of MAC protocols which deal with the conflicting connections in different ways. Here, we shortly describe three *fixed-assignment* strategies used by MAC protocols:

Time Division Multiple Access (TDMA): In this strategy, the shared resource (media channel) is dedicated to each connection in a certain time intervals. As illustration, assume that there are ten different connections which simultaneously need to use a media channel in time interval $[0, 1)$. Then, the ith one can only access the channel in interval $[0, 1/10)$. The same strategy can be periodically repeated for longer time intervals. Note that in this approach, the channel is only accessible by one connection at any given moment.

Frequency Division Multiple Access (FDMA): Every media channel has a bandwidth which can be divided according to the signal frequencies. Regarding our previous example, all the ten connections can simultaneously access the media channel without any conflicts if each one use it in a different frequency interval which is disjoint of the others.

Code Division Multiple Access (CDMA): In this approach, the way of frequency modulation is manipulated in a way that the channel bandwidth gets larger and different sensors can communicate concurrently via the same channel.

Note that we can use a *hybrid* strategy which is a mixture of the above three ones. For example, the channel bandwidth can be divided to some disjoint frequency intervals, in which each interval is used by a number of connections in disjoint time slices.

In addition to the aforementioned *fixed-assignment strategies*, the MAC protocols can be categorized based on the sensor demands. In some strategies, there is a global controller in the DSN which periodically asks every sensor node if it needs the shared resource (*polling*). On the other hand, every sensor node can schedule the earliest available session on demand by sending some reservation messages. This strategy which asks the network to transfers two different types of messages (data and reservation) handles the connection conflicts in a different way (comparing with polling).

Here are some vital concerns which need to be considered regarding the second layer of DSNs because of the existing power restrictions:

- Sensor nodes do not have to *hear* those signals which are not related to them. *Overhearing* is an issue that occurs when an inappropriate MAC protocol is used in DSNs which wastes the power substantially. A possible solution is to define a *sleep mode* for a sensor node in which the node disregards all the messages sent to it.
- In some MAC protocols, some nodes listen to their connecting media channel when no traffic is sending to them. This circumstance which is called *idle listening* may occur in some MAC protocols which consumes the nodes energy for no reason.
- In some MAC protocols which are too complicated, sensor nodes switch to different operational modes frequently which consume the power inefficiently.
- Collision is another issue that results in retransmitting the same packet through the network and wasting the energy.

1.3.3 Network Layer

The middle layer of the network architecture is called "network layer." This layer serves the upper ones as a tool for routing data packets through the network. Every packet which is transmitted through the network at the connection source needs to be forwarded by other sensor nodes to reach its destination. Since every node may have different connections with some specific nodes (its neighbors), the data have to be forwarded elaborately to reach its destination faster and cheaper.

There are many layer-three protocols which can be classified into two different classes based on their architectures: centralized routing protocols and distributed ones. In the first category, one global controller specifies all the data paths; however, in the second one, each sensor node does its duty as a router and forwards the packet based on its *forwarding table*.

The main concern of the network layer protocols in DSNs is to route data in a *congestion-free* manner. In other words, the data flow has to be distributed through the network as the power resource of every sensor node is limited. If the data flow gets congested in some specific nodes, their battery gets discharged frequently which may lead to the network failure. Additionally, the total routing cost needs to be kept small. These concerns urge the network designers to use *oblivious routing algorithms* which balances the traffic load and incurs lower total cost by nature.

1.3.4 Transport Layer

In many real-world examples of DSNs, there exist a group of sensor nodes such that each group has a *base-station*. When a node wants to send some data through the network, it first sends the packet to its nearest base-station using the service provided by the transport layer [13]. Then, the station forwards the packet using the services provided by the first three layers. Additionally, when a packet is received by a base-station, it delivers the data to the appropriate sensor node (destination) using the fourth layer protocol of the network.

Similar to the third layer, load balancing is a critical concern in this layer. High data traffic may lead to the queue overloading at the base-station. This will increase the chance of packet loss and retransmissions which waste the power.

1.3.5 Application Layer

The uppermost layer of DSNs mostly depends on the application we expect. Implementing efficient protocols leads to low-power consumption which is absolutely vital is DSNs.

1.4 Coverage and Connectivity Problems

Before continuing our discussion regarding the coverage and connectivity, some preliminaries are presented.

1.4.1 Preliminaries

Undirected graph G is defined as the ordered pair (V, E) such that V and E are the set of vertices and edges in G, respectively. Moreover, for every edge e in E, $e = \{u, v\}$ where u and v are two *distinct* vertices which are members of V. Edge e is called the connecting edge of vertices u and v if $e = \{u, v\}$.

For some undirected graph, a *simple path* is inductively defined in the following form:

Definition 1.1 In the undirected graph $G = (V, E)$, set $p \subseteq E$ is called a (simple) path from $s \in V$ to $t \in V$ if:

$$p = \begin{cases} \emptyset & s = t \\ \{\{s, v\}\} \cup p' & \text{otherwise} \end{cases}$$

where $\{s, v\} \in E$ undirected graph G is connected if for every pair of vertices $u, v \in V$, there is a path from u to v.

In Definition 1.1, the undirected graph was introduced. As an extension, we define the *weighted* undirected graph.

Definition 1.2 Let $G' = (V, E)$ denote an undirected graph. The triple $G = (V, E, w)$ is defined as the weighted undirected graph corresponding to G', such that V, E, and $w{:}E \mapsto \mathbb{R}^+$ are the vertex set, edge set, and the weight function of G, respectively. Moreover, the (simple) path between two vertices in G is defined as the path between them in G'. Also, G is connected if and only if G' is connected.

Here, we assume that any undirected graph (V, E) is equivalent to the weighted undirected graph $(V, E, unit)$, where $\forall e \in E{:}unit(e) = 1$. In other words, every definition or claim about the weighted undirected graph can be extended to the unweighted graph using this equivalence. Note that in all of the following definitions, weighted undirected graphs and (unweighted) undirected graphs are simply referred to weighted graphs and (unweighted) graphs, respectively.

Considering $G = (V, E, w)$ as a weighted graph, the length of simple path p in G is specified by function $leng_G{:}2^E \mapsto \mathbb{R}^+$ and is defined in the following equation:

$$leng_G(p) = \sum_{e \in p} w(e)$$

Distance function $d_G{:}V^2 \mapsto \mathbb{R}^+$ of weighted graph $G = (V, E, w)$ is defined in the following form:

$$d_G(u, v) = \min_{p \in P(u,v)} leng_G(p) \tag{1.2}$$

where $P(u, v)$ denotes the set of all the simple paths existing between vertices u and v in G. Note that if weighted graph G is replaced with its unweighted equivalent, i.e., if we assume that the weight function of graph G is the *unit* function, the length of a path is obtained by the following equation:

$$leng_G(p) = \sum_{e \in p} unit(e) = \sum_{e \in p} 1 = |p|$$

Hence, the distance between any two vertices of a connected graph is the length of the shortest path existing between them. Additionally, if graph G is not connected, as set $P(u, v)$ will be empty for some $u, v \in V$, we need to redefine the distance function:

$$d_G(u, v) = \begin{cases} \min\limits_{p \in P(u,v)} leng_G(p) & P(u, v) \neq \emptyset \\ +\infty & \text{otherwise} \end{cases}$$

The maximum values of all the finite distances in a connected graph is called the *diameter of graph G* and is represented by $diam(G)$:

$$diam(G) = \begin{cases} \max\limits_{u,v \in V} d_G(u,v) & G \text{ is connected} \\ \text{undefined} & \text{otherwise} \end{cases}$$

Moreover, *an r-neighborhood of vertex* v or *ball* $B_G(v, r)$ in the weighted graph $G = (V, E, w)$ is defined by the following set:

$$B_G(v, r) = \{u \in V | d_G(u, v) \leq r\}$$

For any set $V' \subseteq V$, the *subgraph* of graph $G = (V, E, w)$ *induced by* V' is graph $G_{V'} = (V', E', w)$ where:

$$E' = \Big\{\{u, v\} | u, v \in V' \wedge \{u, v\} \in E\Big\}$$

The triple (V, E, loc) is an unweighted geometric graph in the n-dimensional Euclidean space $\mathcal{S}$ if (V, E) specifies an unweighted graph; and loc denotes a function in the form loc:$V \mapsto \mathcal{S}$. For every (unweighted) geomtric graph, we define three types of distances for each pair of vertices. Assuming that u and v are two vertices of graph $G = (V, E, \text{loc})$, this is the case that:

- The *graph distance* between u and v is denoted by $d_G(u, v)$ and defined as $d_{G'}(u, v)$ where $G' = (V, E)$ represents the unweighted graph corresponding to G. Moreover, if p denotes a path in G, its length $len_G(p)$ is defined as $len_{G'}(p)$.
- The (Euclidean) distance between u and v is denoted by $d_{\text{loc}}(u, v)$ and defined as $||\text{loc}(v) - \text{loc}(u)||$ ($||X - Y||$ represents the Euclidean distance[2] between points X and Y in space $\mathcal{S}$).
- The *hop-by-hop distance* between u and v is represented by $\delta_G(u, v)$ and defined inductively in the following equation:

$$\delta_G(u, v) = \begin{cases} 0 & u = v \\ \min\limits_{\{w,v\} \in E} \{\delta_G(u, w) + ||\text{loc}(w) - \text{loc}(v)||\} & \text{otherwise} \end{cases}$$

Regarding the above-defined distances in a geometric graph, it is inferred by the triangular inequality that for every pair of vertices u and v, the Euclidean distance $d_{\text{loc}}(u, v)$ is not greater than $\delta_G(u, v)$. Also, the *pseudo-diameter* Ψ of geometric graph $G = (V, E, \text{loc})$ is defined in the following way:

$$\Psi(G) = \max_{u \in V} \max_{v \in V} \frac{d_G(u, v)}{d_{\text{loc}}(u, v)}$$

For every point $X \in \mathbb{R}^n$ and value $r \in \mathbb{R}_{\geq 0}$, the n-dimensional ball $B(X, r) \subseteq \mathbb{R}^n$ is defined by the following equation:

[2] Euclidean distance between points $X \in \mathcal{S}$ and $Y \in \mathcal{S}$ is denoted by $||X - Y||$ and defined as the length of line segment $\overline{XY}$: $||X - Y|| = |\overline{XY}|$.

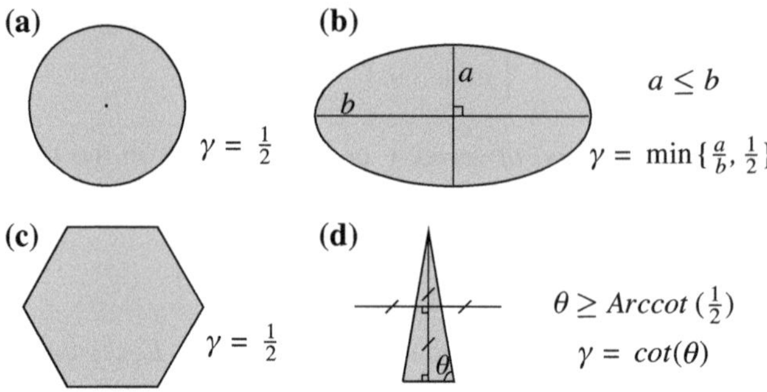

Fig. 1.1 Pseudo-convexity of some familiar convex sets. **a** A circle of unit radius. **b** An ellipse of diameters a and b. **c** A regular hexagon. **d** A bilateral triangle

$$B(X, r) = \left\{Y \in \mathbb{R}^n \big| ||X - Y|| \leq r\right\}$$

The n-dimensional space $\mathcal{S}$ is called to be *thoroughly* $\mathcal{R}$-covered by geometric graph $G = (V, E, loc)$ if this is the case that:

$$\mathcal{S} \subseteq \bigcup_{v \in V} B(\text{loc}(v), \mathcal{R})$$

For every convex subset in the Euclidean plane, a number is defined as its *pseudo-convexity factor* which is formally defined in Definition 1.3.

Definition 1.3 If $\mathcal{S}$ denotes a convex set of points on the Euclidean plane, the *pseudo-convexity factor* $\gamma^{\mathcal{S}}$ is defined as:

$$\gamma^{\mathcal{S}} = \inf_{A, B \in \mathcal{S}} \frac{|\overline{AB}^{\perp} \cap \mathcal{S}|}{|\overline{AB}|}$$

such that line segment $\overline{AB}^{\perp}$ denotes the perpendicular bisecting line segment[3] of $\overline{AB}$; and $|\overline{AB}^{\perp} \cap \mathcal{S}|$ is the length of the part of line segment $\overline{AB}^{\perp}$ that is inside set $\mathcal{S}$.

In the above definition, note that as A and B belong to the convex set $\mathcal{S}$, the result of expression $\overline{AB}^{\perp} \cap \mathcal{S}$ is always a line segment or a point (line segment of length zero). In Fig. 1.1, you can see the pseudo-convexity factor of some familiar convex sets in the Euclidean plane. In this section, assume that γ represents the pseudo-convexity factor of convex area $\mathcal{A}$ over which the routing nodes are distributed.

[3] Line segment $\overline{AB}^{\perp}$ is perpendicular bisecting line segment of $\overline{AB}$, if and only if $\overline{AB}^{\perp} \perp \overline{AB}$, $\overline{AB}$ bisects $\overline{AB}^{\perp}$, and $\overline{AB}^{\perp}$ bisects $\overline{AB}$.

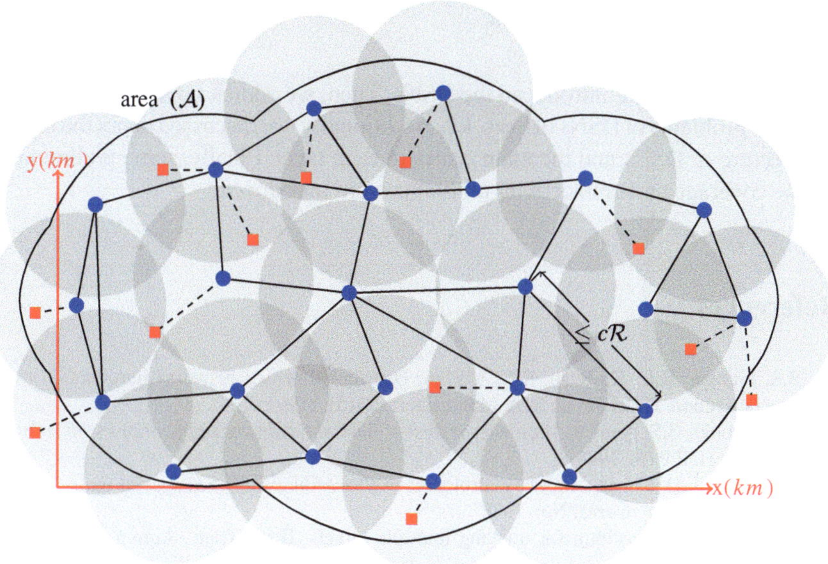

Fig. 1.2 Deployment of the sensor nodes (*blue circles*) and some sensing targets (*red rectangles*) on the Euclidean plane. Note that the radius of any *gray circle* is $\mathcal{R}$. As you see, the sensor nodes thoroughly $\mathcal{R}$-cover area $\mathcal{A}$

1.4.2 Definition

A DSN can be represented as a geometric graph such that every sensor node is denoted by a vertex and the connection between two nodes is represented by the graph edge between corresponding vertices. The vertex location is equal to the coordinates of where the sensor is located in the Euclidean plane/3-D space.

Coverage problem is a fundamental issue in DSNs. In this problem, the goal is to determine how well a set of sensors can monitor a given area [14, 15]. The Art Gallery problem which is closely related to the coverage in DSNs will be addressed in Chap. 2.

The connectivity problem in DSNs is to determine whether the graph representation of the network is connected or not.

Figure 1.2 illustrates the graph representation of a DSN and some target places which need to be sensed by the sensor nodes. In this network, every sensor node has the identical *sensing range*[4] of $\mathcal{R}$ and *communication range*[5] of $c\mathcal{R}$.

We will address the coverage problems in the second part (Chaps. 4 and 5).

[4] The maximum Euclidean distance between the sensor location and the target which is sensed by the sensor.

[5] The maximum Euclidean distance between two adjacent sensor nodes.

1.5 Summary and Outlook

In the first section, we introduced the DSNs. Then, we addressed some famous optimization problems in DSNs in Sect. 1.2. Additionally, we discussed about the layered architecture of DSNs and the main issues in each layer. Finally, some preliminaries for the coverage and connectivity problems were introduced.

References

1. S.S. Iyengar, R.L. Kashyap, R.N. Madan, Distributed sensor networks-introduction to the special section, IEEE Trans. Syst. Man Cybern. **21**(5), (1991)
2. R.R. Brooks, S.S. Iyengar, *Multi-sensor fusion: fundamentals and applications with software*. (Prentice Hall PTR, 1998)
3. R.F. Sprouil, D. Cohen, High level protocols in *Proceedings of IEEE, special issue on packet communication networks*, Nov 1978
4. D.B. Reid, An algorithm for tracking multiple targets, IEEE Trans. Autom. Contr. **AC-24** (1979)
5. S. Mori, C.Y. Chong, R.P. Wishner, E. Tse, Multi-target multi-sensor tracing problems: a general bayesian approach, in *Proceedings of American Control Conference*. (San Francisco, CA, 1983)
6. C.Y. Chong, S. Mori, Hierarchical multi-target tracking and classification—a Baysian approach, in *Proceedings of American Control Conference*. (San Diego, CA, 1984)
7. C.Y. Chong, E. Tse, S. Mori, Distributed estimation in networks, in *Proceedings of American Control Conference*. (San Francisco, CA, 1983)
8. C.Y. Chong, et al., Distributed hypothesis testing in distributed sensor networks, Artificial Intelligence & DS, Tech. Rep TP-1048-02, July 1984
9. J. O'Rourke, Art gallery theorems and algorithms. (Oxford University Press, 1987)
10. E.H. Callaway, *Wireless Sensor Networks: Architectures and Protocols*. (Auerbach Publications, New York, 2003)
11. H. Karl, A. Willig, *Protocols and Architectures for Wireless Sensor Networks*. (Wiley, New York, 2005)
12. I.F. Akyildiz, W. Su, Y. Sankarasubramaniam, E. Cayirci, Wireless sensor networks: a survey. Comput. Netw. **38**, 393–422 (2002)
13. A. Rahman, A. El Saddik, W. Gueaieb, Wireless Sensor Network Transport Layer: State of the Art, vol. 21, Sensors, *Lecture Notes in Electrical Engineering*. (Springer, Berlin, 2008), pp. 221–245
14. S. Kumar, T.H. Lai, and A. Arora, Barrier coverage with wireless sensor networks, in *Proceedings of the 11th Annual International Conference on Mobile Computing and Networking*. (Cologne, Germany, 2005), pp. 284–298
15. D.W. Gage, Command control for many-robot systems in *Proceedings of Nineteenth Annual AUVS Technical Symposium*, 1992, pp. 22–24

Part I
Optimization Methods

Chapter 2
Region-Guarding Problem in 3-D Areas

This chapter studies the optimal inspection of autonomous robots in a complex pipeline system. We solve a 3-D region-guarding problem to suggest the necessary inspection spots. The proposed hierarchical integer linear programming (HILP) optimization algorithm seeks the fewest spots necessary to cover the entire given 3-D region. Unlike most existing pipeline inspection systems that focus on designing mobility and control of the explore robots, this chapter focuses on global planning of the thorough and automatic inspection of a complex environment. We demonstrate the efficacy of the computation framework using a simulated environment, where scanned pipelines and existing leaks, clogs, and deformation can be thoroughly detected by an autonomous prototype robot.

2.1 Introduction

Active monitoring and frequent inspections are critical to maintaining pipeline health. As the most economical way to transport gas, oil, biofuels, water resource, sewer, and so forth, pipelines have become an indispensable part of our daily lives. However, pipelines always suffer from aging and damages, which can cause great waste of resource, environmental pollution, and many other incidence. For example, the leak of petroleum pipeline causes ocean pollution and ecocatastrophe. Regular inspections and maintenance of pipelines are essential to keep them functional.

Unfortunately, the difficulty and the cost for human inspection can be extremely high, especially with the appearance of increasingly complicated pipelines nowadays (see Fig. 2.1 for illustration). There are several reasons:

This chapter has been reprinted with permission from "On Optimizing Autonomous Pipeline Inspection," Xin Li, Member, IEEE, Wuyi Yu, Student Member, IEEE, Xiao Lin, and S. S. Iyengar, Fellow, IEEE, IEEE Transactions on Robotics, Vol. 28, No. 1, February 2012.

DOI: 10.1007/978-1-4419-8420-3_2,

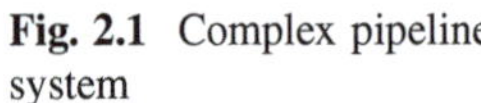

Fig. 2.1 Complex pipeline system

- Pipeline systems are often buried/hided underground or into walls. Hiding pipelines' presence from the surrounding environment is necessary for better protection of pipelines, as well as the elegance of the architecture.
- The structure of pipelines is usually designed long, thin, and complex, in order to conduct long-distance transportation or circumvent-complicated terrain or limited space of the architecture structure.
- The environment inside pipelines can be dirty and hazardous: Sewerage water or hazardous gas can be overflowing.

These factors make direct artificial inspection oftentimes prohibitive and, therefore, significantly increase the costs for the maintenance of pipelines. For example, to inspect the pipelines, people can dig holes in different pipeline sections for the inspection; high professional competence could be necessary especially when the environment condition is severe. Indirect methods include placing sensors outside the pipeline and monitoring parameters such as pressure and temperature. However, the sensibility is easily affected by environment and the material transmitting in the pipeline. Pipeline clogs are sometimes directly penetrated with long sticks or wires, but it can be very difficult if pipes are curved or circumvented; another common approach is to blow out blockages using air pressure, which fails, however, if pipes have multiple outlets or cracks. In any case, to apply repair, the suspected area for clogging or leaking needs to be located. This step usually takes the longest time and largest cost.

2.1.1 Pipeline Inspection by Autonomous Robots

With the development of autonomous robots and imaginary sensing technologies, pipeline robots that are equipped with cameras and sensors become ideal candidates to avoid tedious artificial inspection for automatic pipeline inspection and repair.

Current robotic inspection systems (see Sect. 2.2) usually have a robot that is equipped with a camera and sensors; the robot moves around and transmits the captured images back to a remote monitor for the operator's inspection. The robot's movement and the camera direction need to be manually controlled by a skillful inspector. Such an interactive monitoring system provides easier and safer inspection. However, it can still be *labor intensive (costly)* and *time consuming*. In addition, complex pipeline environments could limit the extensive use of these remotely controlled pipeline robots. The mobility of remotely controlled robots could not meet the requirement for the inspection in complex pipeline environments, such as the urban gas pipeline and chemical pipelines. The wired connections also limit the operation range of robots. Furthermore, the thoroughness of the examination may not be guaranteed and heavily relies on expertise of the operator.

An autonomous robot that can routinely inspect the environment and report cracks, clogs, or deformation will, therefore, be highly desirable. If such an inspection system can be developed reliably and conducted routinely, it will greatly save artificial costs and prevent the abnormal situations in important pipelines. To develop an efficient inspection plan for robots, according to different environments, to ensure inspection reliability (thoroughness) is related to several challenging geometric problems.

2.1.2 Optimal Autonomous Inspection by Region Guarding

Naturally, one wants to ask the following fundamental open problem for autonomous inspection: How do we conduct the most efficient yet thorough inspection? More specifically, given an environment to inspect, how many inspection spots are necessary to visually cover the entire region? The solution directly dictates the correctness and efficiency of an autonomous inspection system and, therefore, is critical.

Visually covering a given 3-D region is an interesting geometric problem called gallery guarding, defined (see Sect. 2.3 for details) as follows: Given a region whose boundary is a surface, find the smallest set of points inside the region from which all the boundary points (i.e., the wall) are visible. This problem, having high complexity, has been actively studied in 2-D. In general, 2-D polygonal regions, this problem has been proved to be NP-hard. To the best of our knowledge, on 3-D regions approximated by polyhedra, which have much higher complexity, this problem is little explored and no efficient algorithm has been reported.

In this chapter, we design a HILP algorithm to find an approximate optimal solution for the guarding of a given 3-D region. Compared with the greedy and the optimal algorithms, HILP has a good approximation to the optimal solution (for instance, to guard a region in Fig. 2.3d, an optimal guarding needs 13 guards, the greedy approach needs 18 points, while our HILP algorithm guards it with 14 points) but is several orders of magnitude less than the direct optimization on time complexity.

Effective region guarding can greatly benefit the automatic pipeline inspection. Compared with existing manual inspection systems, the biggest advantage of the new inspection system built upon optimal guarding is its thorough (therefore, making

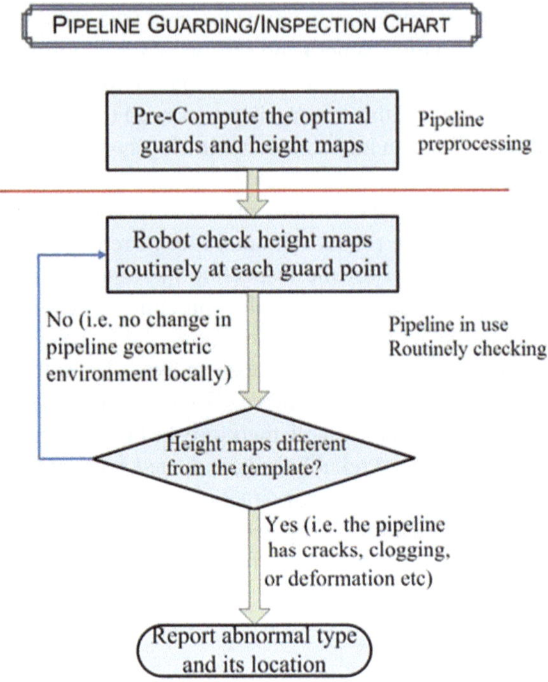

Fig. 2.2 Autonomous inspection based on region guarding

the system robust) inspection using fewest (therefore, making the system efficient and inexpensive) necessary checking spots. The proposed inspection framework is illustrated in Fig. 2.2, and the pipeline has two steps.

1. *Preprocessing Stage*: Once pipeline is newly installed or when it is working well, we compute its optimal guarding, e.g., a small set of points $\{g_i\}$. By checking on these spots, the entire pipeline can be visually covered. Then, a pipeline robot only goes to these points, scans the pipe, and builds up sequential height maps as templates. These height maps characterize the original pipeline geometry.
2. *Online Stage*: The robot will go into the pipeline to conduct inspection routinely. Every time, it only needs to move to these spots $\{g_i\}$, scan the depth information of the surrounding environment, and compare these height maps with the corresponding templates. Abnormal geometry changes such as cracks, clogs, and deformation can be detected and located immediately.

The main contributions of this work lie in both efficiently finding good approximate solutions of the NP-hard 3-D guarding problem and its application on robotic inspection.

1. *Optimality*: We develop an efficient algorithm to find approximate solution to 3-D region guarding. The solution indicates a smallest set of spots from which thorough inspection can be most timely and costly efficient.

2. *Autonomy*: We design an automatic pipeline inspection system for autonomous robots. Unlike other existing systems whose inspection quality heavily relies on manual controls, in our framework, robots can inspect on these fewest guarding points autonomously yet thoroughly.
3. *Generality and Robustness*: The algorithm efficacy is demonstrated in our simulated platform. It is also generally applicable on various robot systems, such as our pipe robot prototype FAMPER [16], which is equipped with a range sensor that provides 2.5-D range image with depth information. Furthermore, combined with 2-D image reconstruction techniques, our algorithm can work well for robots that are equipped with a conventional 2-D camera.

2.2 Background and Related Work

2.2.1 Gallery Guarding

On a geometric region M, we want to find the optimal guarding, which uses the smallest number of points $\{g_i\}$ inside M so that any boundary point $p \in \partial M$ is visible to at least one guard. Here, M is a 3-D shape whose boundary ∂M is represented by a polygonal mesh. For a given guard g_i, any point $p \in \partial M$ is visible to g_i if the line segment $\overline{g_i p}$ is entirely located inside M (we consider $\partial M \subseteq M$). Various versions of this problem are generally called art gallery problems which are known to be a famous problem with high complexity. Even in the 2-D case, the problem is known to be NP-complete. "Very little is known about gallery guarding in three dimensions" [24]. To our best knowledge, no effective approximation algorithm has been proposed for 3-D regions that are bounded by general polygons, and this is the first practical algorithm that works for large free-form 3-D domains (such as complicated pipeline systems) represented by polygonal meshes.

The art gallery problem was first proposed by Klee. Guards can be restricted to boundary vertices ($p \in \partial M$), interior vertices ($p \in M$), or mobile vertices. When guards are not mobile, they are called stationary guards. If guards are restricted to the boundary, they are called vertex guards; if there is no boundary restriction, the guards are referred as point guards. In 2-D, Chavatal [2] and Fisk [13] both showed that a simple polygon $M \subset \mathbb{R}^2$ needs at most $\lfloor n/3 \rfloor$ stationary guards, based on which Avis and Toussaint [12] developed an $O(n \log n)$ time algorithm to position $\lfloor n/3 \rfloor$ guards in M. When guards are mobile, we call them mobile guards. Furthermore, mobile guards are called edge guards if they are restricted to boundary vertices. O'Rourke [5] showed that $\lfloor n/4 \rfloor$ mobile guards are sufficient for a simple polygon $M \subset \mathbb{R}^2$. More results are recapped in Table 2.1.

The aforementioned theoretic work discusses the conservative upper bounds for necessary guards on various regions. Given a specific region, we are interested in designing practical algorithm to find its optimal point guards, which depends on topology and geometry of this region. An effective algorithm to compute the optimal

Table 2.1 Upper bounds for gallery-guarding problem

Work	M type	Guard type	Optimal bound
Chvatal [5]	Simple, 2D	Stationary	$\lfloor \frac{n}{3} \rfloor$
Rourke [36, 37]	Simple, 2D	Mobile	$\lfloor \frac{n}{4} \rfloor$
Urrutia [44]	Simple, 2D	Edge	$\lfloor \frac{n}{4} \rfloor$
Kahn et al. [19] and Rourke [36, 37]	Orthogonal, 2D	Stationary	$\lfloor \frac{n}{4} \rfloor$
Rourke [35]	h holes, 2D	Vertex	$\lfloor \frac{n+2h}{3} \rfloor$
Hoffmann et al. [13] and Bjorling-Sachs et al. [3]	h holes, 2D	Point	$\lfloor \frac{n+h}{3} \rfloor$
Gyori et al. [24]	Orthogonal, h holes, 2D	Mobile	$\lfloor \frac{3n+4h+4}{16} \rfloor$

guarding of a given region will benefit many geometric computing tasks. However, computation of optimal guarding is highly challenging. Finding minimal guards has been shown to be NP-hard for 2-D polygons with holes [1], 2-D simple polygons [34], and even 2-D simple orthogonal polygons [27, 39], using either vertex or point guards. Approximation algorithms have been studied in 2-D to get a close to optimal result in polynomial time complexity. Ben-Moshe et al. [24] cover 1.5-D terrain using point guards in $O(n^2)$ time, with the optimal factor $O(1)$. Efrat and Har-Peled [21] find vertex guards for 2-D simple polygonal regions and h-hole polygonal regions in $O(nc_{\text{opt}}^2 \log^4 n)$ and $O(nc_{\text{opt}}^2 \log^4 n)$ expected time, with expected $O(\log c_{\text{opt}})$ and $O(nhc_{\text{opt}^3} \log^4 n)$ optimal factors, respectively. Lien [10] computes guarding for 3-D point cloud data, approximating visibility using ϵ-view. The algorithm is based on a randomized greedy approach.

2.2.2 Pipeline Inspection Robots

According to the degree of autonomy, pipeline inspection robots can be classified as follows [30].

1. *No Autonomy*: Robots are fully tele-operated by humans via a tether cable. While the robot is traveling through the pipe, the pipeline condition data are collected and sent back by the robot and then assessed by human operators.
2. *Semiautonomy*: Robots are partially controlled by automatic control programs [33].
3. *Full Autonomy*: Robots are fully controlled by programs and perform an automatic pipeline condition assessment. However, lacking effective technologies in efficient analysis of environment hinders the automatic assessment [30].

2.2.2.1 Manual and Semi-Autonomous Methods

There are many application inspection technologies, such as closed-circuit television (CCTV), laser surveys, sonar surveys [4], radio frequency identification [7], and mobile sensor [22]. The introduction of CCTV inspection methods in the 1960s provided an inexpensive and safe option, and they, thus, have been the most popular and widely used approaches across the industry for many years. The CCTV provides rich videos/images information, which is collected by robots for subsequent pipeline condition assessment. Various CCTV methods have similar principles. The robot is mounted by a remotely controlled tractor, carries a television camera, and illuminates the interior of the pipe. The inspector has to identify and categorize defects by the image displayed on the monitor. When a defect is noticed, the inspector stops the robot and assesses the condition.

Advances in optical survey techniques have been utilized in the sewer scanner and evaluation technology (SSET) such as in [23]. Unlike the CCTV inspection system, the SSET may not need to stop for a zooming-in defect inspection. For instance, recent work in [45] has advanced the use of automated defect detection systems for pipelines.

Laser-based systems and ultrasonic-based systems are also used in pipeline inspection (see the survey in [25]). Laser-based systems are generally implemented in two ways: the whole-circle image method and the single-spot scanning method [9]. The first method projects a full ring of light onto the wall in one go, while the single-spot scanning method sends point-by-point beams in sequential. These two methods indicate a trade-off between accuracy and inspection time. The whole-ring image method allows faster data acquisition but has been found less accurate [46]. Ultrasonic-based systems use high-frequency sound waves to detect pipe properties such as thickness, shape, and presence/sizes of defects [32]. Laser-based and ultrasonic-based methods can be combined to obtain higher quality data [40].

2.2.2.2 Autonomous Methods

A few full-autonomous robots have been developed for pipeline inspection. The Kurt [28] can run in dry clean pipelines guided by maps uploaded into the robot. The Marko [15] is designed for autonomous navigation in clean pipelines with diameter ranges from 300 to 600 mm. The Kantaro [30] is used to navigate in pipelines with diameter ranges from 200 to 300 mm, but only the horizontal mobility is considered. To design autonomous pipeline inspection robots, the main challenges include their moving ability, energy, and pipe condition assessment [41].

In this chapter, we use laser range finders to detect the depth (height) information toward sets of sample directions. The captured 3-D range images provide easy measurement of the environment. We focus on designing the algorithm and architecture of the effective autonomous inspection system. The guarding and subsequent inspection can easily extend to various systems that are based on different data acquisition schemes.

2.3 3-D Gallery Guarding

The geometric abnormalities of the pipeline can be detected from the robot if the robot can see this region and measure the distance from itself to the pipe. Suppose the robot always checks at a set of same spots, and it has premeasured (template) distance information on each spot toward different directions, then it can tell whether the current pipeline is normal, i.e., preserving the same shape. These checking spots need to be intelligently selected so that fewest comparisons are necessary. Meanwhile, to guarantee that the entire pipeline is visually covered, we require that these points together can guard the entire region.

Given a point p inside the region M, suppose we represent the boundary surface of M using a triangle mesh $\partial M = T, V$, where $V = \{v_1, v_2, \ldots, v_{N_V}\}$ is the vertex set, and $T = \{t1, t2, \ldots, t_{N_T}\}$ is the set of triangles connecting them. We say that p is visible to a point q on ∂M (q is not necessary a vertex; it can be a point on a triangle from T), if the line segment $\overline{pq}$ connecting p and q is totally inside M, namely $\overline{pq}$ intersects ∂M only on q. We call the set of all visible points on the boundary $\{q\}$, $q \in \partial M$ the visible region $S(p)$ of p. Then, we say that a set of points $\{p\}$ can visibly guard the entire region, if the union of their visible regions is the entire ∂M. Finding a smallest guarding set $\{p\}$ that can cover the entire region is the optimal guarding problem that we want to solve. Our algorithm is based on the following intuitions.

1. As demonstrated in several medical visualization and virtual navigation applications (e.g., [14, 31]), medial axes (curve skeletons) usually have desirable visibility to boundary points (referred as the "reliability" of skeletons). An effective skeleton can guide the camera navigation, ensuring nice examination (visibly covered) of the interior of organ surfaces.
2. Hierarchical skeletons or skeletons for a progressively simplified mesh can be effectively computed and used to reduce the size of the optimization problem, leading to a computation of better numerical efficiency and stability against boundary perturbations.

Many effective skeletonization algorithms (see a survey by Cornea et al. [20]) have been developed for 3-D shapes. We use the algorithm in [6] since it efficiently generates skeletons on medial axis surfaces of the 3-D shapes. Suppose the boundary surface ∂M of a volumetric region M is represented by a triangle mesh (also denoted as ∂M) with n vertices, and the output skeleton has k nodes; the guarding problem is then converted to finding a minimal-size point set G from this k points, such that all n boundary vertices are visible to G.

2.3.1 Visibility Detection

A basic operation is to detect the visible region $S(p)$ of a given point p. Following the definition, for a point $q \in \partial M$, to check its visibility to a point $p \in M$, one should check intersection between the line segment $\overline{pq}$ and ∂M. If the intersection is detected on a point $q' \in \partial M$ other than q and the Euclidean distance $|\overline{pq'}| < |\overline{pq}|$, then q is not visible from p. The intersection between the line segment $\overline{pq}$ and ∂M can be detected by checking the intersection between $\overline{pq}$ and each triangle face $t_i \in T \subseteq \partial M$.

We need to compute the visibility of p against all the vertices of the mesh of ∂M. Simply enumerating every $\overline{pv_i}$ to check its intersections with every triangle $t \in T$ is time consuming: For a single interior point p, it takes $O(N_V \cdot N_T) = O(N_V^2)$ time to check its visibility on all boundary vertices. We develop the following sweep algorithm to improve the efficiency to $O(N_T \log N_T)$, i.e., $O(N_V \log N_V)$.

We create a spherical coordinate system which is originated at p. Each vertex $v_i \in V$ is represented as $\overline{pv_i} = \big(r(v_i), \theta(v_i), \phi(v_i)\big)$, where $r(v_i) \geq 0$ and this is the case that:

$$\begin{cases} -\pi < \theta(v_i) \leq \pi \\ -\frac{\pi}{2} \leq \phi(v_i) \leq \frac{\pi}{2} \end{cases}$$

For every $i = 1 \ldots N_T$, let t_i denote triangle $\Delta v_{i,1} v_{i,2} v_{i,3} \in T$. We contract the following notations:

$$\begin{cases} \theta_{\max}(t_i) = \max_{j=1}^{3}\{\theta(v_{i,j})\} \\ \theta_{\min}(t_i) = \min_{j=1}^{3}\{\theta(v_{i,j})\} \\ \phi_{\max}(t_i) = \max_{j=1}^{3}\{\phi(v_{i,j})\} \\ \phi_{\min}(t_i) = \min_{j=1}^{3}\{\phi(v_{i,j})\} \end{cases}$$

The segment $\overline{pv_k}$ cannot intersect with a triangle t unless they are adjacent:

$$\begin{cases} \theta_{\min} \leq \theta(v_k) \leq \theta_{\max}(t) \\ \phi_{\min}(t) \leq \phi(v_k) \leq \phi_{\max}(t) \end{cases} \tag{2.1}$$

The angle functions θ and ϕ are not continuously defined on a sphere. When a triangle t spans $\theta = \pi$, we duplicate it to ensure that each θ of the original t is in interval $\big[\theta_{\min}(t) - 2\pi, \theta_{\min}(t)\big)$ and θ of its duplicate is in interval $\big[\theta_{\max}(t) - 2\pi, \theta_{\max}(t)\big)$, by adding or subtracting θ by 2π. For each triangle t that spans $\phi = \pi$, we detect and duplicate it in the same way. Using $\theta(v_i)$ as the primary key and $\phi(v_i)$ as the secondary key, we then sort all line segments $\overline{pv_i}$. Then, we sweep all segments following the angle functions one by one, filtering out triangles not satisfying Condition 2.1. Specifically, we define a counter c_i on every triangle t_i. Initially, $c_i \leftarrow 0$, when the segment $\overline{pv}$ for some $v \in t_i$ is being processed, $c_i \leftarrow c_i + 1$. The following two cases indicate that the sweep has not reached the neighborhood of the triangle t_i, and subsequently, we do not need to check its intersection with line segment $\overline{pv}$.

$$\begin{cases} c_i = 0 \rightarrow \big(\theta_{\min}(t_i) > \theta(\overline{ov})\big) \vee \big(\phi_{\min} > \phi(\overline{ov})\big) \\ c_i > 3 \rightarrow \big(\theta_{\max}(t_i) < \theta(\overline{ov})\big) \vee \big(\phi_{\max} < \phi(\overline{ov})\big) \end{cases} \tag{2.2}$$

Therefore, we maintain a list L of neighboring triangles $\{t_i\}$ whose counters have $c_i \in \{1, 2, 3\}$. When the sweep segment hits a new triangle t_j, we make $c_j \leftarrow 1$ and add t_j into L; when a counter $c_j = 3$, after processing the current segment, we remove t_j from L.

Given a skeleton point p, for a boundary triangle mesh with N_V vertices, it takes $O(N_T \log N_T)$ to compute and sort all triangles following their segment angles. When we are sweeping a segment $\overline{pv_i}$, if the size of the active triangle list L is m, it takes $O(m)$ intersection-detecting operations. Therefore, the total complexity is $O(N_T \log N_T + N_V \cdot m)$. The incident triangle around a vertex v_i is generally very small (i.e., $m < \log N_T$). Therefore, the algorithm finishes visibility detection of p in $O(N_T \log N_T)$ time. On a skeleton containing k nodes, it takes $O(kN_T \log N_T)$ precomputation time to compute the visible region for all nodes.

2.3.2 Greedy and Optimal Guarding

Once visibility information for all skeletal nodes is computed, we want to pick a minimum sized point set that can cover all boundary vertices. It can now be converted to a set-covering problem which is also NP-complete [8]: Given the universe point set $V = \{v_i | i = 1, \ldots, N_V\}$, and a family S of subsets $S_j = \{s_{j,k}\} \subseteq V$ (for every $j = 1, \ldots, N_S$), a *cover* is a subfamily $C \subseteq S$ of sets whose union is V. We want to find a cover C that uses the fewest subsets in S. Here, V corresponds to the set of all vertices of ∂M; for each skeletal node p_j, $j = 1, \ldots, N_S$, S_j contains all the boundary vertices visible to p_j. Each C indicates a subset of skeletal nodes that can guard the entire region. Skeletons generated using medial axis-based methods with dense enough nodes usually ensure that S itself is a covering. This holds in all of our experiments. However, if a coarsely sampled skeleton cannot cover the entire V, we can easily include all those invisible vertices, i.e., their visible regions, into S.

A *greedy* strategy for the set covering is to iteratively pick the skeletal nodes p that can cover the largest number of unguarded vertices in V, then remove all guarded vertices $v \in S(p)$ from V meanwhile, and update S accordingly since the universe becomes smaller, until $V = \emptyset$. The greedy strategy is effective, and it yields $O(\log n)$ approximation [19] to the set-covering problem.

An *optimal* selection can be computed by 0–1 programming, also called integer linear programming (ILP). For every skeleton point p_i, $i = 1, \ldots, N_S$, we assign a variable x_i such that:

$$\begin{cases} 1 & \text{if } p_i \text{ is chosen} \\ 0 & \text{otherwise} \end{cases}$$

The *objective* function to minimize is then $\sum_{i=1}^{N_S} x_i$, as we want to pick the fewest necessary points. Since every vertex $v \in V$ should be covered, for each such v_i, at least one of its visible skeletal nodes $P_i = \{p_j | v_i \in S(p_j)\}$ should be picked. Therefore, we minimize $\sum_{i=1}^{N_S} x_i$ subject to:

$$x_i = \{0, 1\} \wedge \sum_{p_j \in P_i} x_j \geq 1 \quad \forall v_i \in V$$

This objective function can be minimized using branch-and-bound algorithms. When the dimension is small (e.g., a few hundreds to a few thousands), we can use the TomLab optimization package [18] to solve it efficiently.

2.3.3 Hierarchical Guarding

Computing the optimal guarding based on ILP is highly time consuming and it limits the size of problems that we can handle: General 3-D volumetric shapes can easily have a number of vertices (20,000–200,000) on its boundary surface, which is too large for this optimization. On the other hand, the greedy algorithm generates the guards in a locally optimal manner. Furthermore, the greedy strategy is not robust against local geometric perturbations. For example, a small bump could lead to global structural variance of the guarding points. We propose a hierarchical guarding computation framework which is based on the progressive mesh [43], combining the 0–1 programming optimization and the adaptive greedy refinement.

We simplify the boundary mesh ∂M into several resolutions in the following form using a progressive mesh [43].

$$\partial M^i = \{V^i, F^i\} \quad \forall i = 0, \ldots, m$$

In the coarsest level $i = m$, ILP optimization is performed on all elements $v \in \partial M^m$, and we get the coarsest level guard set $G_i = \{g_k^i\}$. Then, we progress to $i = m - 1$ level $\partial M^{m-1} = (V^{m-1}, F^{m-1})$.

1. Map existing guards $G_{i+1} = \{g_k\}$ to the closest finer level skeletal nodes $G_i = \{g_k'\}$ to locally adjust them to maximize their visible region $S(g_k')$.
2. Remove the least significant guards $\left\{g \middle| \, |S(g)| < \varepsilon N_V\right\}$ from G^i.
3. Remove covered vertices $\{v | v \in S(g) \wedge g \in G^i\}$.

Then, we solve ILP again on uncovered boundary vertices. With details increase in finer levels, new guards will be inserted into G^i. Before applying the ILP optimization, we further conduct four *reduction* operations (see in the following) on uncovered regions to reduce the dimensions of the optimization. This progressive refinement ends when all boundary vertices are covered on the finest level $i = 0$.

Reduction: The dimension of the ILP optimization on each level can be reduced using the following reduction rules, without changing the size of the optimal solution. Suppose we store the visibility information in an incidence matrix A. If the skeletal node p_i can see the vertex v_j, then we let $a_{ij} = 1$, otherwise (p_i cannot see v_j), let $a_{ij} = 0$. Originally, the dimension of A is $N_S \times N_V$. The following four rules are applied to reduce it.

1. If column j has only one nonzero element at row i, we must take p_i in order to see v_j. Therefore, add p_i into G and remove column j. In addition, for all nonzero element a_{ik}, remove column k (we take p_i: all points that it sees are guaranteed to be covered, and thus now can be removed).
2. If for the couple of rows i_1 and i_2, the following proposition holds:

$$a_{i_1 j} = 1 \rightarrow a_{i_2 j} = 1 \quad \forall j,$$

then p_{i_2} sees all vertices that p_{i_1} can see; and we can remove the entire row i_1.
3. If for the couple of columns j_1 and j_2, the following proposition holds:

$$a_{i j_1} = 1 \rightarrow a_{i j_2} = 1 \quad \forall i,$$

then guarding v_{j_1} guarantees the guarding of v_{j_2}, and we can remove the entire column j_2.
4. If the matrix A is composed of several blocks, we partition A into several small matrices $\{A_k\}$.

In step 4, after removing vertices that have been seen by the adjusted guards from a coarser level, remaining boundary vertices could be partitioned to several connected components far away from each other, which can be optimized separately and more efficiently.

In our experiments, we simplify the boundary mesh to the coarsest level with 5,000 vertices for the first round ILP optimization. Generally, we make each iteration to add in another 10,000 vertices. When the size of constraints is around 5,000 and the size of variables (skeletal nodes) is around 1,000, the optimization usually takes 10–50 s to solve.

Our hierarchical scheme together with the reduction processing has the following important advantages over both the pure greedy strategy and the pure 0–1 optimization.

1. It is *much faster* than the nonlinear ILP optimization. The current framework can handle large-size geometric shapes.
2. With similar performance, it usually provides *better* guarding solutions than a pure greedy strategy.
3. It is *hierarchical* and therefore is *robust* and *stable* against geometric noise. In our HILP framework, refined local details tend to not change the global structure of the previously optimized guarding graph in coarser levels.

Figure 2.3 shows some examples of HILP guarding on sculpture data, and Fig. 2.1 shows the run-time statistics. We use these irregular sculpture data to demonstrate the significant efficacy of our algorithm since our prototype pipelines on hand are relatively simple. To thoroughly cover and inspect complex pipeline system such as in Fig. 2.1, using HILP, we can find its guarding point set efficiently.

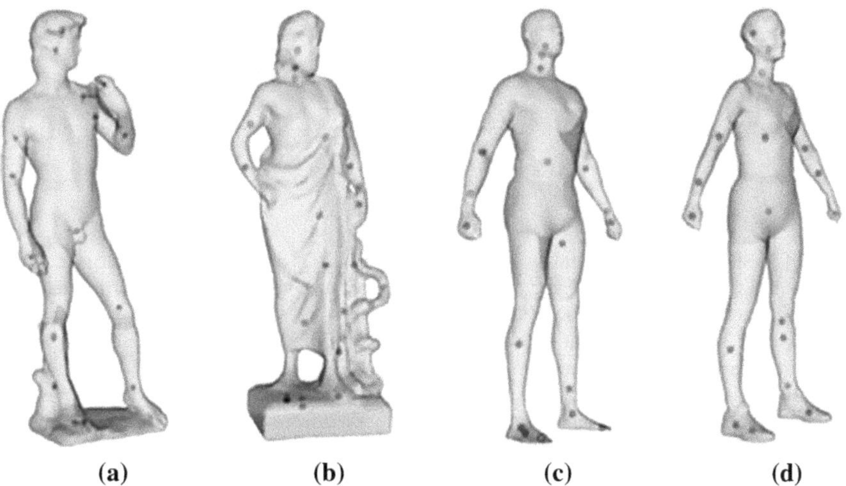

Fig. 2.3 Guarding statues using HILP. From *left* to *right* guarding on a few sculpture datasets. **a** Michelangelo's David. **b** Greek. **c** Cyberware Male. **d** Female. Small nodes are the guards, where *green nodes* are the latest computed guards on the finest level

2.4 Autonomous Pipeline Inspection

Our guarding algorithm computes the set of necessary checking spots for thorough inspection. The inspection robot only needs to go to each guarding point, construct the current height maps (see Sect. 4.4.1), and compare them with the precalculated templates for abnormal identifications. When geometric changes are detected, the system refines the identification of the abnormal areas (see Sect. 4.4.2) and extracts the boundary of the damaged region (see Sect. 4.4.3).

2.4.1 Height Maps Acquisition

The geometry of the pipeline environment is measured using laser range finders in our system. The distance from the robot to a point on the wall is captured and stored. A range image stores a set of depth information in a rectangle view-port along a direction, and we call it a height map.

The abnormal detection is based on the comparison between precalculated height map templates and the current height maps. We compare height maps on each inspection point. Suppose the laser scanning has two parameters, the scan range angle α and the sampling rate $s \cdot \alpha$ decide the field of view of the scanner, and s indicates the sampling resolution inside the field of view. As Fig. 2.4a shows, given a shooting direction, the scanner takes a snapshot of the environment, which produces a height map on a planar square region R, uniformly sampled with $s \times s$ points P. Since

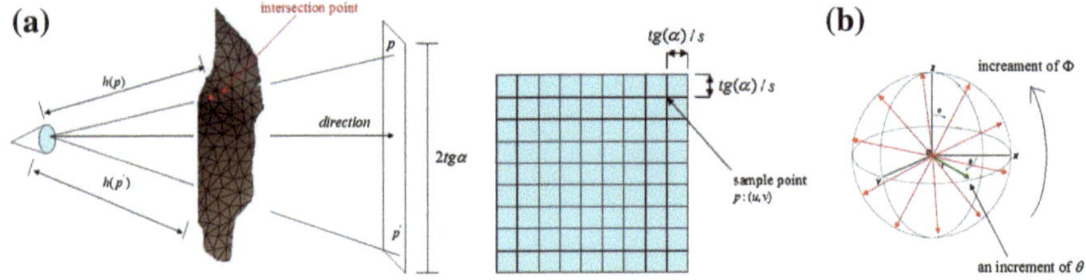

Fig. 2.4 **a** Shot of the laser scanning. The sample grid density is decided by the sample rate and the scan range of the laser scanner. Every sample point has a planar coordinate (u, v) corresponding to the actual intersection point in 3-D space. The laser scanner rotates to a direction and does a snapshot to project the surface to a planar with a range of $2\tan\alpha$. The *red* point indicates the intersection result of p in the 3-D region, and the corresponding height data $h(p)$ are stored for detection. Scanning at every sample point, the height map is constructed. **b** Motion plan of the scanner. The scanner rotates to cover the whole spherical area. The *red arrowheads* are the rotated direction of ϕ, and the *green* is one step of rotation increment of θ

one snapshot can only cover an area within the current view angle, a planed motion sequence is necessary for the laser scanner to rotate and take pictures to cover the entire 360°. The laser scanner with the scan range α and the sample rate s is placed at a guarding position o which points toward an initial direction L; the height map acquisition processes can be simply conducted as follows.

1. Using the local spherical coordinate system which is defined at o, given a direction $L(r, \phi, \theta)$, L takes a snapshot and gets a depth image $P(\alpha, s)$ (see Fig. 2.4a). Depth is defined on every point on the image $p \in P$, whose 2-D coordinate can be defined as (u, v), for every $(u, v) \in [0, 1]^2$.
2. 3-D position of each boundary point v can be derived from the depth on its projection $p(u, v)$. The transformation can be represented by a rotation matrix.
3. After taking one depth image, the laser scanner rotates for another height map. The motion plan of the scanner follows the rotation sequences: first fix ϕ, rotate θ for N times, increase $\pi\alpha$ iteratively, and then increase ϕ to perform the θ rotation again until the whole spherical region is covered. The rotation path is shown in Fig. 2.4b.
4. Finally, we get the set of height maps $\{H_L\}$ collected in the aforementioned steps and save them together with the starting projection direction on every guarding point.

In practice, during height map acquisition, the geometry of regions far away from a guard can be visible but captured less accurately because of the precision of the range finder or the sampling resolution. To tackle this issue, before guarding computation, we add a simple parameter d_p for each skeleton node p: If a boundary point $q \in \partial M$ is visible to p but far away, i.e., distance $|\overline{pq}| > d_p$, we consider q to be invisible. In other words, a guard only sees points in a bounded distance. The whole optimization algorithm can be applied exactly in the same way. In long and thin environments, more guards may be necessary, but each guarding region will have less long antenna, and the inspection accuracy will be improved. A heuristic

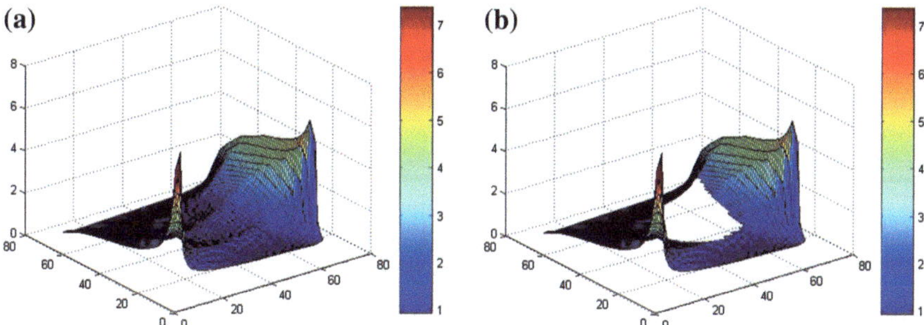

Fig. 2.5 Height maps comparison. **a** Template height map. **b** Height map of an abnormal region. The missing region indicates a hole

setting for d_p can be $d_p = \alpha \cdot d_p^0$, where α is a constant parameter, and d_p^0 is the distance from p to its nearest visible boundary point.

2.4.2 Abnormal Boundary Detection

After all height maps are obtained, on every guarding point, we compare the precollected template height maps $\{H_L\}$ (see Fig. 2.5a) and the current height maps $\{H_L'\}$ (see Fig. 2.5b). If the height information changes; i.e., $|h(u,v) - h'(u,v)| > \varepsilon$, we consider the region around (u,v) as a defective region and report (u,v) as a abnormal point. ε is a similarity threshold. In practice, the acquired depth data could have geometric and topological noise. By adjusting ε, the system can tolerate small deviations because of certain acquisition noise. On the other hand, a local efficient data preprocessing step in topological denoise or geometric completion/fairing [17, 38] could also be helpful in cleaning environment noise.

Compared with reporting simply a set of sampled abnormal points, an accurate estimation of the bad region's shape is desirable. When a defective region is detected, in order to "zoom in" to see the shape of the defective region, we need to examine more sampling points by performing denser scanning around this region. During the initial scanning of the original pipeline, we may not do very dense sampling; therefore, with the same number of depth acquisition, we can capture a larger region in every shot for better efficiency. Regions among sampling points are approximated using bilinear interpolation $h(E)$:

$$h(E) \approx \frac{h(u_1,v_1)}{(u_2-u_1)(v_2-v_1)}(u_2-u)(v_2-v) + \frac{h(u_2,v_1)}{(u_2-u_1)(v_2-v_1)}(u-u_1)(v_2-v) + \frac{h(u_1,v_2)}{(u_2-u_1)(v_2-v_1)}(u_2-u)(v-v_1) + \frac{h(u_2,v_2)}{(u_2-u_1)(v_2-v_1)}(u-u_1)(v-v_1)$$

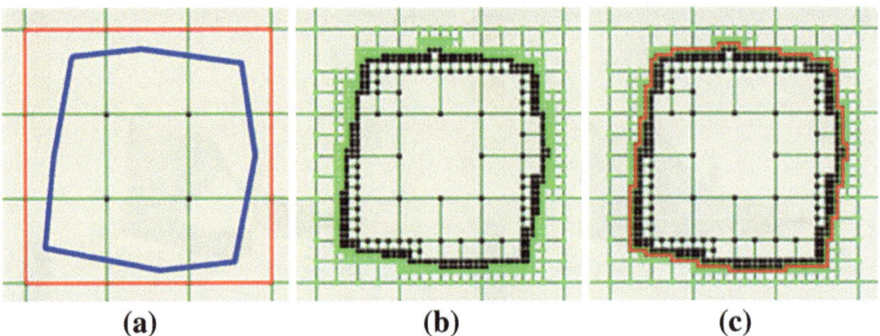

Fig. 2.6 Quadtree splitting and boundary extraction. **a** *Polygon colored in blue* indicates a defective region on the pipe wall. The *green nodes* are normal points, and the *black* ones indicate the abnormal height data; the boundary extracted from the quadtree without quadtree splitting, colored in *red*, is coarse. **b** Quadtree cells split. **c** Extracted boundary (*red curves*) after quadtree refinement

To report a refined boundary shape of an abnormal region, we use a quadtree [29] subdivision scheme: Near such a region, boundary cells split accordingly to get the refined geometry. For each new point (u', v'), we detect its height $h'(u', v')$ and compare it with the interpolated height on the template height map $h(u', vv')$. Figure 2.5 shows an example. The max level of the subdivision is determined by the resolution of the laser range finder. The boundary of holes, clogs, and deformations can be detected/refined using this paradigm, since these geometric changes always lead to changed height maps.

2.4.3 Boundary Extraction

When the refinement is done, the boundary of the abnormal region can be extracted and reported. We conservatively link the normal points on boundary cell to form the boundary: Starting from a normal point on a boundary cell, linking the normal points in the cell along the edge, getting a neighboring boundary cell, and repeating this process until all boundary cells are traversed.

Figure 2.6 illustrates the process of quadtree splitting and extraction. In a snapshot, the grids are the cells, and the green polygon indicates a hole on the pipeline wall. Normal points are colored in green, while abnormal points are colored in black. In the coarsest resolution, we get the red boundary shown in (a). We can get finer boundary shape (b) and the extracted boundary is depicted in red in (c).

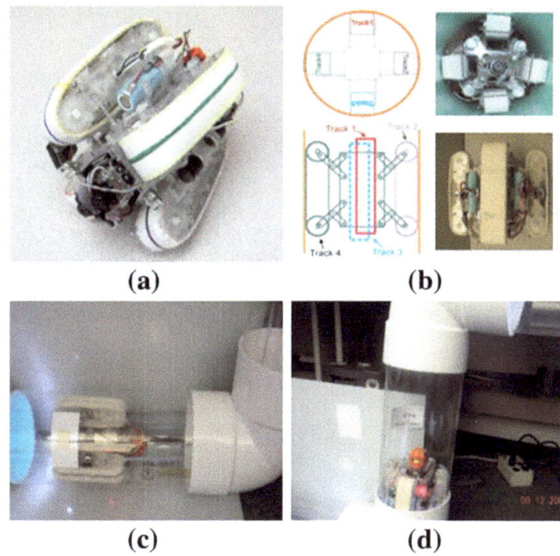

Fig. 2.7 Prototype robot FAMPER. **a**, **b** Robot and its design. **c**, **d** Robot inspecting a pipe

2.5 Result of Simulated Experiments

Our guarding-driven pipe inspection system can be implemented on the prototype pipeline robot [16], which can be used for the inspection of pipelines. This robot consists of four wall-press caterpillars that are operated by two dc motors each to provide steering capability to go through 45°, 90° elbows, *T*-branches, and *Y*-branches and make a superior performance in all types of complex networks of pipelines (see Fig. 2.7). The robot is also equipped with a powerful computing system that makes it extendable to various sensing and actuating devices, such as in our experimental system, for localization and laser scanning. The height information can be obtained using a multislit laser range scanner [11], which has the size of about 110 m × 90 mm and includes a laser projector and a charge-coupled device (CCD) camera (the laser projector in [11] is StockerYale Mini-715L, which projects 15 slits, with an adjacent slit angle 2.3°; the CCD camera is Point Grey Research Flea2, which measures 330 points). The height information can also be obtained by laser scanners in [15, 30] or other range cameras (e.g., [26, 42]).

The rotation of the sensor can be controlled by a small mechanical platform installed in the rear part of the robot; therefore, the scan can be conducted radially inside the pipe toward different directions. Several effective sensing platforms with similar mechanism have been developed [30].

We develop a simulated platform for testing our algorithm (see Fig. 2.8), which simulates the process of inspection. We apply our procedure on this platform using several complicated 3-D virtual pipelines. The simulated results are convincing and show the effective inspection on pipeline geometry.

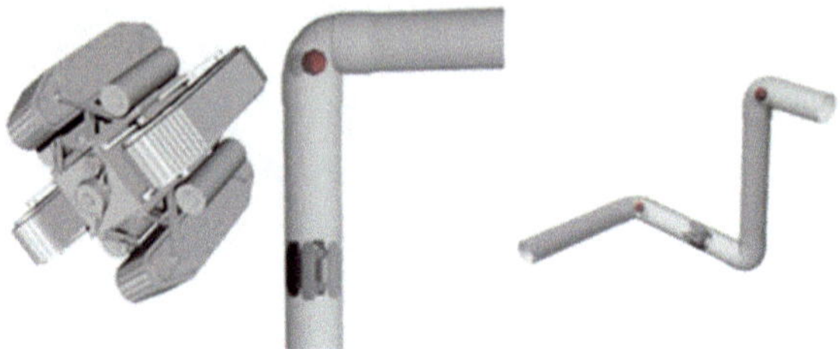

Fig. 2.8 Simulation environment. The 3-D model of the FAMPHER robot in the pipeline; *red points* denote guarding points

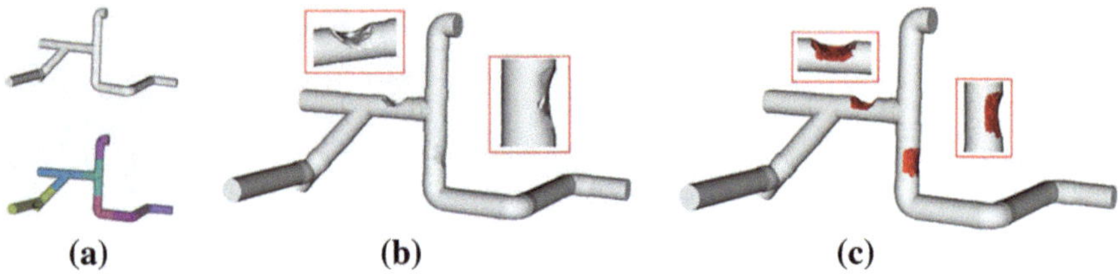

Fig. 2.9 Inspection process on a simulated pipeline. **a** *Upper* simulated model. *Lower* regions guarded by different guard points encoded in different colors. **b** Damaged pipe with some holes. **c** Damages are detected, whose boundaries are extracted and shown in *green*

2.5.1 Hole Detection

If a hole appears on the pipeline, it can be identified online when the robot reaches the guarding point that covers this region and matches the captured range depth images with the stored templates. We simulate this on pipeline meshes M by randomly generating some missing regions. An experiment is shown in Fig. 2.9. A pipeline model and the necessary guards are shown in (a), where regions covered by each guard are rendered in a specific color for the visualization purpose. Any given region of the pipeline is covered by at least one guard and, therefore, is colorized. The height maps can then be generated as templates, measuring the "correct" distance from each guarding site to the pipe wall toward specific directions. This simulates range images obtained by a laser scanner. Now, we simulate the appearance of defected regions on the pipeline by generating some missing regions as shown in (b). When the robot checks height maps on guarding points, these holes can be immediately detected and illustrated in (c).

Another example is shown in Fig. 2.10; this pipe is guarded by 12 points (a). In addition, the damaged region of the pipe is big and with complex topology (b). In this case, the robot should check from more than one guarding points in order to detect the entire shape of such a big hole. The entire defect geometry is extracted by composing boundaries detected from different guarding sites. The merged boundary loop is illustrated in (c).

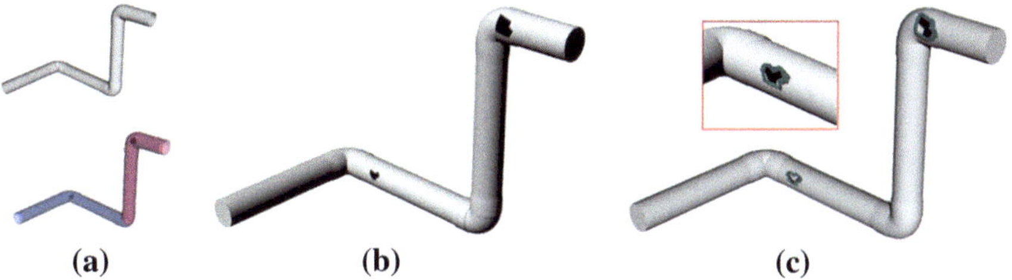

Fig. 2.10 Guarding and inspection process on a more complicated pipeline. **a** *Upper* simulated model. *Lower* guarding points and guarded regions rendered in different colors, respectively. **b** Damaged pipe with big and concave holes. **c** Large holes are detected/extracted from more than one guarding points; the whole big boundary is composed of several extracted subboundaries and identified separately from different guarding points, as shown in *green*

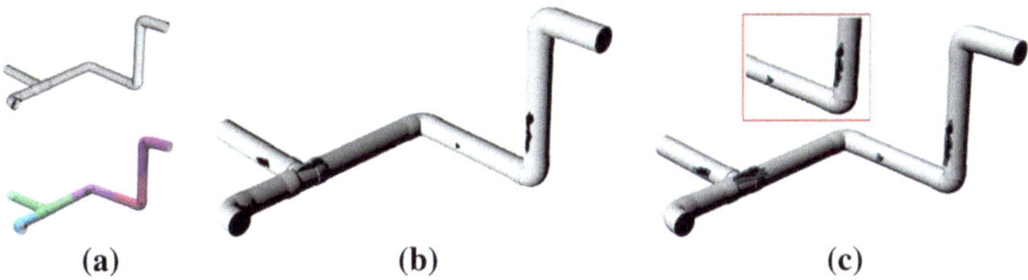

Fig. 2.11 Inspection process on a deformed pipeline. **a** *Upper* simulated model. *Lower* regions guarded by different guard points encoded in different colors. **b** Deformed pipe. **c** Deformations are detected, whose regions are extracted (and refined) and shown in *red*

2.5.2 Deformation Detection

Small deformation such as bending and erosion can also be detected in our system as shown in Fig. 2.11. The detected deformed region is colored in red.

2.5.3 Clogging Detection

Clogging also changes the scanned geometry of the pipeline and can be detected. Figure 2.12 shows an example. The clogged solid (green) is detected, and its boundary geometry is reconstructed using height maps as illustrated in dark red. The robot will report the clogs when it is detected. In this example, the reconstruction merges the geometry of the blocking stuff from two aspects (from two guarding points) using their corresponding height maps.

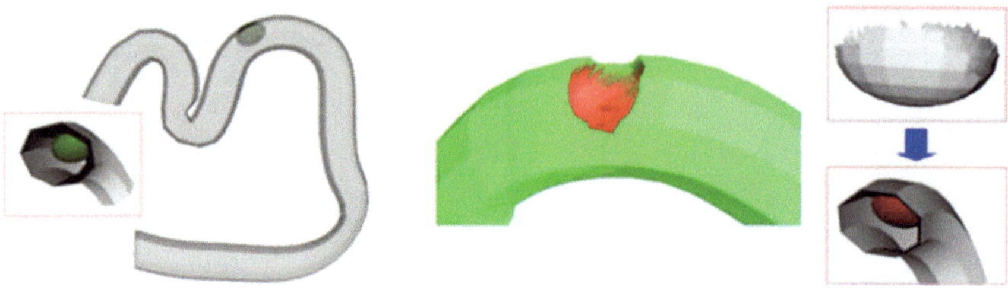

Fig. 2.12 *Left* pipeline blocked by a solid (*in green*). *Middle* clog detection on the pipeline. Height maps constructed and merged by detected views from different guarding points. *Green* regions indicate normal status, while the *red* regions indicate the abnormal height information corresponding to the clog geometry. *Right* inspection result, which is the reconstructed surface colored in *dark red*, describes the location and geometry of the blocking region

2.6 Summary and Outlook

We have proposed an efficient 3-D guarding algorithm that can cover a given complicated environment using as few as possible points. Finding an efficient solution to the fundamental problem of how to inspect on fewest points yet thoroughly covering the entire environment can greatly benefit the autonomous design of inspection and exploration robots. We have developed a simulation system of pipeline inspection and conducted experiments to evaluate the efficacy of our system. With our optimal guarding, abnormal geometric changes of the pipeline such as holes, clogs, and deformation can be thoroughly detected online. The remaining challenging issue for the current system is the dynamic environment mapping. The initial geometry of the pipeline system needs to be scanned to a digital model before the region-guarding computation. In addition, after the guarding spots are computed, the pipeline wall needs to be marked so that the robot can localize itself to know whether it is on the spot. The development of a system that a robot without this prior knowledge can do simultaneous 3-D mapping (i.e., reconstructing the map of the environment) and localization will be highly desirable. In the near future, we will study these localization and dynamic environment mapping problems by exploring effective partial matching of range images.

References

1. D. Avis, G. Toussaint, An efficient algorithm for decomposing a polygon into star-shaped polygons. Pattern Recogn. **13**, 395–398 (1981)
2. B. Ben-Moshe, M.J. Katz, J.S. Mitchell, A constant-factor approximation algorithm for optimal terrain guarding, in *Proceedings of ACM/SIAM Symposium on Discrete Algorithms*, (2005), pp. 515–524
3. I. Bjorling-Sachs, D. Souvaine, An efficient algorithm for guard placement in polygons with holes. Discr. Comput. Geom. **13**, 77–109 (1995)

4. J.Y. Choi, H. Lim, B.-J. Yi, Semi-automatic pipeline inspection robot systems, in *Proceedings of SICE-ICASE International Joint Conference*, (2006), pp. 2266–2269
5. V. Chvatal, A combinatorial theorem in plane geometry. J. Comb. Theory **18**, 39–41 (1975)
6. N. Cornea, D. Silver, P. Min, Curve-skeleton properties, applications, and algorithms. IEEE Trans. Vis. Comput. Graph. **13**(3), 530–548 (2007)
7. S. Costello, D. Chapman, C. Rogers, N. Metje, Underground asset location and condition assessment technologies. Tunn. Undergr. Space Technol. **22**(5–6), 524–542 (2007)
8. T.K. Dey, J. Sun, Defining and computing curve-skeletons with medial geodesic function, in *Proceedings of 4th Eurographics Symposium on Geometry Processing*, (Eurographics Association, Aire-la-Ville, Switzerland, 2006) pp. 143–152
9. O. Duran, K. Althoefer, L. Seneviratne, State of the art in sensor technologies for sewer inspection. IEEE Sens. J. **2**(2), 73–81 (2002)
10. A. Efrat, S. Har-Peled, Guarding galleries and terrains. Inf. Process. Lett. **100**(6), 238–245 (2006)
11. R. Finkel, J. Bentley, Quad trees: a data structure for retrieval on composite keys. Acta Inf. **4**, 1–9 (1974)
12. S. Fisk, Ashort proof of Chvatal's watchman theorem. J. Comb. Theory Ser. B **24**(3), 374 (1978)
13. E. Gyori, F. Hoffmann, K. Kriegel, T. Shermer, Generalized guarding and partitioning for rectilinear polygons. Comput. Geom. **6**(1), 21–44 (1996)
14. T. He, L. Hong, D. Chen, Z. Liang, Reliable path for virtual endoscopy: ensuring complete examination of human organs. IEEE Trans. Vis. Comput. Graph. **7**(4), 333–342 (2001)
15. J. Hertzberg, F. Kirchner, Landmark-based autonomous navigation in sewerage pipes, in *Proceedings of 1st Euromicro Workshop on Advance Mobile Robot*, (Oct 1996), pp. 68–73
16. Hogeschoolzeeland, (2007), Pipes, tubes, machinery and steam turbine at a power plant [Online]. Available http://goo.gl/fRp3c
17. H. Hoppe, Progressive meshes, in *Proceedings of SIGGRAPH*, (New York, 1996), pp. 99–108
18. D.S. Johnson, Approximation algorithms for combinatorial problems, in *Proceedings of 5th Annual ACM Symposium on Theory of Computing*, (New York, 1973), pp. 38–49
19. J. Kahn, M. Klawe, D. Kleitman, Traditional galleries require fewer watchmen. SIAM J. Algebraic Discr. Methods **4**(2), 194–206 (1983)
20. D.-G. Kang, J.B. Ra, A new path planning algorithm for maximizing visibility in computed tomography colonography. IEEE Trans. Med. Imag. **24**(8), 957–968 (2005)
21. M.J. Katz, G.S. Roisman, On guarding the vertices of rectilinear domains. Comput. Geom.: Theory Appl. 39(3), 219–228 (2008)
22. J.-H. Kim, G. Sharma, N. Boudriga, S. Iyengar, Ramp system for proactive pipeline monitoring, in *Proceedings of International Conference on Communication System, Network*, (2010), pp. 1–2
23. J. Kim, G. Sharma, N. Boudriga, S. Iyengar, Spamms:Asensor-based pipeline autonomous monitoring and maintenance system, in *Proceedings of International Conference on Communication System, Network*, (2010), pp. 1–10
24. J.-H. Kim, G. Sharma, S. Iyengar, FAMPER: a fully autonomous mobile robot for pipeline exploration, in *Proceedings of 2010 IEEE International Conference on Industrial Technology*, (2010), pp. 517–523
25. D.-H. Koo, S.T. Ariaratnam, Innovative method for assessment of underground sewer pipe condition. Autom. Construct. **15**(4), 479–488 (2006)
26. T. Kuroki, K. Terabayashi, K. Umeda, Construction of a compact range image sensor using multi-slit laser projector and obstacle detection of a humanoidwith the sensor, in *Proceedings of IEEE/RSJ International Conference on Intelligent Robots and Systems*, (Oct 2010), pp. 5972–5977
27. D.T. Lee, A.K. Lin, Computational complexity of art gallery problems. IEEE Trans. Inf. Theory **32**(2), 276–282 (1986)
28. D. Levesque, M. Ochiai, A. Blouin, R. Talbot, A. Fukumoto, J.-P. Monchalin, Laser-ultrasonic inspection of surface-breaking tight cracks in metals using SAFT processing. Proc. IEEE Ultrasound Symp. **1**(8–11), 753–756 (2002)

29. X. Li, Z. Yin, L. Wei, S. Wan, W. Yu, M. Li, Symmetry and template guided completion of damaged skulls. Comput. Graph. **35**, 885–893 (2011)
30. J.-M. Lien, *Approximate star-shaped decomposition of point set data, in Eurographics Symposium on Point-Based Graphics* (Czech Republic, Prague, Presented at the Eurograph, 2007)
31. J.M. Mirats Tur, W. Garthwaite, Robotic devices for water main in-pipe inspection: a survey. J. Field Robot. 27, 491–508 (2010)
32. J. Moraleda, A. Ollero, M. Orte, A robotic system for internal inspection of water pipelines. IEEE Robot. Autom. Mag. **6**(3), 30–41 (1999)
33. A. Nassiraei, Y. Kawamura, A. Ahrary, Y. Mikuriya, K. Ishii, Concept and design of a fully autonomous sewer pipe inspection mobile robot "KANTARO", in *Proceedings of IEEE International Conference on Robotics and Automation*, 10–14 Apr 2007, pp. 136–143
34. J. O'Rourke, K. Supowit, Some NP-hard polygon decomposition problems. IEEE Trans. Inf. Theory **29**, 181–190 (1983)
35. J. O'Rourke, *Art Gallery Theorems and Algorithms* (Oxford University Press, London, 1987)
36. J.O. Rourke, Galleries need fewer mobile guards: a variation on Chvatal's theorem. Geom. Dedicata **14**, 273–283 (1983)
37. J.O. Rourke, An alternate proof of the rectilinear art gallery theorem. J. Geom. **21**, 118–130 (1983)
38. H. Schoner, B. Moser, A.A. Dorrington, A.D. Payne, M.J. Cree, B. Heise, F. Bauer, A clustering based denoising technique for range images of time of flight cameras, in *Proceedings of International Conference on Computational Intelligence for Modelling Control and Automation*, (2008), pp. 999–1004
39. D. Schuchardt, H.-D. Hecker, Two np-hard art-gallery problems for ortho-polygons. Math. Logic. Quart. **41**, 261–267 (1995)
40. M. Silk, The determination of crack penetration using ultrasonic surface waves. NDT Int. **9**(6), 290–297 (1976)
41. H. Streich, O. Adria, Software approach for the autonomous inspection robot MAKRO. *Proceedings of IEEE International Conference on Robotics and Automation* , vol. 4, pp. 3411–3416 (Apr. 2004)
42. J. Thielemann, G. Breivik, A. Berge, Pipeline landmark detection for autonomous robot navigation using time-of-flight imagery, in *Proceedings of IEEE Computer Society Conference on Computer Vision and Pattern Recognition Workshops*, (23–28 Jun 2008), pp. 1–7
43. TOMLAB v3.0 User's Guide, Technical report IMa-TOM-2001-01, Department of Mathematics and Physics, Malardalen University, Vasteras, Sweden, 2001
44. J. Urrutia, *Art gallery and illumination problems, in Handbook of Computational Geometry, Amsterdam* (North-Holland, TheNetherlands, 2000)
45. R. Wirahadikusumah, D.M. Abraham, T. Iseley, R.K. Prasanth, Assessment technologies for sewer system rehabilitation. Autom. Construct. **7**(4), 259–270 (1998)
46. W.W. Zhang, B.H. Zhuang, Non-contact laser inspection for the inner wall surface of a pipe. Meas. Sci. Technol. **9**(9), 1380 (1998)

Chapter 3
Expectation–Maximization for Acoustic Source Localization

Wideband source localization using acoustic sensor networks has been drawing a lot of research interest recently. The maximum likelihood (ML) is the predominant objective which leads to a variety of source localization approaches. However, the robust and efficient optimization algorithms are still being pursuit by researchers since different aspects about the effectiveness of such algorithms have to be addressed on different circumstances. In this chapter, we would like to combat the source localization based on the realistic assumption where the sources are corrupted by the noises with nonuniform variances. We focus on the two popular source localization methods for solving this problem, namely the stepwise-concentrated maximum-likelihood (SC-ML) and approximately concentrated maximum-likelihood (AC-ML) algorithms. We explore the respective limitations of these two methods and design a new expectation–maximization (EM) algorithm. Furthermore, we provide the Cramer–Rao lower bound (CRLB) for all these three methods. Through Monte Carlo simulations, we demonstrate that our proposed EM algorithm outperforms the SC-ML and AC-ML methods in terms of the localization accuracy, and the root-mean-square (RMS) error of our EM algorithm is closer to the derived CRLB than both SC-ML and AC-ML methods.

An erratum to this chapter is available at 10.1007/978-1-4419-8420-3_7

DOI: 10.1007/978-1-4419-8420-3_3, © Springer Science+Business Media New York 2014

3.1 Introduction

Localization using low-cost and low-complexity sensor arrays has been the active research area in the fields of radar, sonar, geophysics, wireless systems, and acoustic tracking for years [1, 2]. Recently, the wideband source localization in the near field has drawn a lot of research interest in the signal processing applications [3–6]. Extensive studies for the wideband source localization can be found in [3, 4]. Among them, the ML approach in [3] has been regarded as the optimal and robust scheme for coherent source signals. However, when the multiple sources are present, the ML approach facilitates a nonlinear optimization problem, which is impractical especially for the energy-constrained sensor networks. In addition, many of the existing ML estimators are based on the unrealistic spatially white noise (SWN) assumption across different sensors [5–7], where the noise process at each sensor is assumed to be spatially uncorrelated white Gaussian with an identical variance. It is shown that under this assumption, the ML estimates of the unknown parameters (source waveforms/spectra and noise variance) can be expressed as the respective functions of the source locations, and the number of independent parameters to be estimated is greatly reduced. Thus, this assumption, although unrealistic, substantially reduces the search space and usually leads to more efficient localization algorithms. Hence, various wideband ML source location estimators were proposed in [3].

However, this SWN assumption is unrealistic in many applications. In several practical applications [7], the sensors are sparsely placed so that the sensor noise processes are spatially uncorrelated. However, the noise variance of each sensor can still be quite different due to either the variation of the manufacturing process, the imperfection of the sensor array calibration or the "unquiet" background. As a result, the spatial noise covariance matrix (across the sensors) can be modeled as a diagonal matrix where the diagonal elements in general are not identical. Note that this noise model is definitely not a special case of the ARMA model as was explained in [8]. Furthermore, the source location estimators derived from the SWN assumption would often not provide satisfactory results in the real environment since the algorithms derived from the SWN assumption blindly treat all sensors equally in the estimated likelihood. Motivated by the arguments above, in [7], two DOA calculation algorithms, namely SC-ML estimator and AC-ML algorithm, have been recently proposed for the multiple wideband sources. Although both SC-ML and AC-ML methods can be extended for the source localization, the robustness issue still remain challenging in this research area. This is the primary reason why we would like to dedicate this chapter to address these two issues by designing a new source localization scheme.

Felder and Weinstein proposed the generic expectation–maximization (EM) algorithm in [9] to estimate the parameters associated with the superimposed signals and employed it for the array signal processing in [10]. EM-based techniques have also been applied for the multisensor signal enhancement [11–13]. In addition, EM-based narrowband source localization algorithms were proposed by [14, 15]. In this chapter, we modify the EM algorithm to tackle with the general multiple source localization problem when the wideband sources are present in the near field, which evolves from the simple DOA estimation method for the narrowband sources in the far field in [16]. If the wideband sources are considered, the source signal signature or characteristics are unavailable at the sensor array, and the method in [14–16] cannot be applied according to [17, 18]. Therefore, similar to [17, 18], we use the discrete Fourier transform (DFT) filter bank to decompose the wideband signals collected by the sensors and then estimate the complete set of parameters involving source waveforms (or spectra) and source locations. Note that our previous works in [17, 18] can only deal with the source localization problem under the unrealistic SWN assumption. In this chapter, we reformulate the source localization problem for the realistic SNWN assumption and design a new EM-based localization algorithm for multiple wideband sources, and it can be shown that our proposed algorithm is much more computationally efficient and robust than the existing SC-ML and AC-ML methods (we have extended the original SC-ML and AC-ML methods in [7] which could only solve the DOA problem in [8] to combat the source localization problem).

The rest of this chapter is organized as follows. The problem formulation and the signal model are introduced in Sect. 3.2. The ML source location estimators for both SWN and SNWN models are introduced in Sect. 3.3. The novel EM algorithm for wideband source localization in the near field under the SNWN assumption is derived and discussed in Sect. 3.4. In addition, the Cramer–Rao lower bound (CRLB) derivation will be manifested in Sect. 3.5. Monte Carlo simulation results for demonstrating our proposed new EM method and illustrating our newly derived robustness analysis will be provided in Sect. 3.6. Conclusion will be drawn in Sect. 3.7.

Nomenclatures: The sets of all real and complex numbers are denoted by $\mathbb{R}$ and $\mathbb{C}$, respectively. A vector is denoted by $\underline{A}$ and a matrix is denoted by $\widetilde{A}$. The statistical expectation operation is expressed as $E[\,]$. Besides, $\widetilde{A}^T$, $\widetilde{A}^*$, $\widetilde{A}^H$, $\det(\widetilde{A})$, $\widetilde{A}^\dagger$, and $\mathrm{trace}(\widetilde{A})$ stand for the transpose, conjugate, Hermitian adjoint, determinant, pseudo-inverse, and trace of the matrix, respectively. In addition, $\odot$ stands for the Hadamard matrix product operator, and $\|\,\|$ stands for the Euclidean norm.

3.2 Signal Model

According to [3], we consider a randomly distributed array of P sensors to collect the data from M sources. Since the sources are assumed to be in the near field, the signal gains are different across the sensors. Thus, the signal collected by the th sensor at a discrete time instant n is given by:

$$x_p(n) = \sum_{m=1}^{M} a_p^{(m)} s^{(m)}\left(n - t_p^{(m)}\right) + w_p(n) \tag{3.1}$$

for $n = 0, \ldots, L-1$, $p = 1 \ldots, P$, $m = 1, \ldots, M$, where $a_p^{(m)}$ is the gain of the mth source signal arriving at the pth sensor; $s^{(m)}(n)$ denotes the mth source signal waveform; $t_p^{(m)}$ is the propagation delay (in data samples) incurred from the mth source to the pth sensor; $w_p(n)$ represents the zero mean independently identically distributed (i.i.d.) noise process. Several crucial parameters are specified as follows:

- $t_p^{(m)}$ represents the propagation delay from the mth source to the pth sensor:

$$t_p^{(m)} \stackrel{\text{def}}{=} \frac{F_s \left(||\underline{r_s}^{(m)} - \underline{r_p}|| \right)}{v}$$

- $\underline{r_s}^{(m)} \in \mathbb{R}^{2\times 1}$ denotes the mth source location,
- $\underline{r_p} \in \mathbb{R}^{2\times 1}$ represents the pth sensor location,
- v denotes the source signal propagation speed in meters/second,
- F_s represents the sampling frequency.

Taking the N-point DFT of both sides in Eq. 3.1 and reserving a half of them due to the symmetry property, we have:

$$\underline{X}(k) = \widetilde{D}(k)\underline{S}(k) + \underline{U}(k) \quad \forall k = 0, \ldots, \frac{N}{2} - 1 \tag{3.2}$$

where

$$\underline{X}(k) \stackrel{\text{def}}{=} [X_1(k), \ldots, X_P(k)]^T \in \mathbb{C}^{P\times 1} \tag{3.3}$$

and $X_p(k)$ is the kth DFT point of $x_p(n)$, for every $p = 1, \ldots, P$. The symbols for the right-hand side of Eq. 3.2 are clarified as follows:

$$\widetilde{D}(k) \stackrel{\text{def}}{=} \left[\underline{d}^{(1)}(k), \ldots, \underline{d}^{(M)}(k)\right] \in \mathbb{C}^{P\times M} \tag{3.4}$$

consists of M *steering vectors*, each given by:

$$\underline{d}^{(m)}(k) \stackrel{\text{def}}{=} \left[d_1^{(m)}(k), \ldots, d_P^{(m)}(k)\right]^T \in \mathbb{C}^{P\times 1} \quad \forall m = 1, \ldots, M \tag{3.5}$$

where

$$d_p^{(m)}(k) \stackrel{\text{def}}{=} a_p^{(m)} e^{-\frac{j2\pi k t_p^{(m)}}{N}} \tag{3.6}$$

and $j \stackrel{\text{def}}{=} \sqrt{-1}$. Note that:

$$\underline{S}(k) \stackrel{\text{def}}{=} \left[S^{(1)}(k), \ldots, S^{(M)}(k)\right]^T \in \mathbb{C}^{M\times 1} \tag{3.7}$$

consists of M individual source signal spectra, each given by $S^{(m)}(k)$ where $S^{(m)}(k)$ is the kth DFT point of $s^{(m)}(n)$, for every $m = 1, \ldots, M$.

In reality, the source signal spectral vector $\underline{S}(k)$ is unknown and deterministic. The noise spectral vector is $\underline{U}(k) \in \mathbb{C}^{P \times 1}$ is a complex-valued zero-mean spatially uncorrelated Gaussian process with the following covariance matrix:

$$\widetilde{Q} \stackrel{\text{def}}{=} E\left[\underline{U}(k)\underline{U}(k)^H\right] = \begin{bmatrix} q_1 & 0 & \cdots & 0 \\ 0 & q_2 & \cdots & 0 \\ \vdots & \vdots & \ddots & \vdots \\ 0 & 0 & \cdots & q_P \end{bmatrix} \in \mathbb{C}^{P \times P} \quad \forall k \tag{3.8}$$

In general, $q_p s$ (for every $p = 1, \ldots, P$) are not necessarily identical to each other under the SNWN assumption. Hence, we need to deal with the realistic source localization problem in the presence of the nonuniform noise variances thereupon.

3.3 Maximum Likelihood for Source Localization

Prior to the establishment of the log-likelihood for the source localization in the presence of the nonuniform noise variances as stated by Eq. 3.8, we start from the conventional ML formulation for the identical noise variance across the sensors.

3.3.1 Conventional Maximum Likelihood for Source Localization in the Presence of Identical Noise Variance (SWN)

According to the signal model given by Eq. 3.2 together with the noise variance constraint as $\widetilde{Q} = \sigma^2 \widetilde{I}$, where σ^2 is the noise variance and $\widetilde{I}$ is a $P \times P$ identity matrix, the ML source localization formulation can be facilitated as [1, 3, 7]. We highlight the relevant pivotal formula here.

Let $\underline{r_s}$, $\widetilde{S}$, and σ^2 represent all the unknown parameters in Eq. 3.2 necessary to be estimated, where:

$$\underline{r_s} \stackrel{\text{def}}{=} \left[\underline{r_s}^{(1)^T}, \ldots, \underline{r_s}^{(m)^T}, \ldots, \underline{r_s}^{(M)^T}\right]^T \in \mathbb{R}^{2M \times 1} \tag{3.9}$$

and

$$\widetilde{S} \stackrel{\text{def}}{=} \left[\underline{S}(0)^T, \ldots, \underline{S}\left(\frac{N}{2} - 1\right)^T\right]^T \in \mathbb{C}^{(\frac{NM}{2}) \times 1} \tag{3.10}$$

In addition, we denote the *residual vector* as:

$$\underline{g}(k) \stackrel{\text{def}}{=} [g_1(k), \ldots, g_P(k)]^T = \underline{X}(k) - \widetilde{D}(k)\underline{S}(k) \in \mathbb{C}^{P \times 1} \tag{3.11}$$

Thus, the likelihood function is given by

$$f(\underline{r_s}, \widetilde{S}, \sigma^2) \stackrel{\text{def}}{=} \frac{1}{\pi^{\frac{PN}{2}} \sigma^{PN}} \exp\left\{ -\frac{1}{\sigma^2} \sum_{k=0}^{\frac{N}{2}-1} \underline{g}(k)^2 \right\} \tag{3.12}$$

Taking the logarithm of Eq. 3.12 and neglecting all the constant terms, we can derive the log-likelihood function $L(\underline{r_s}, \widetilde{S}, \sigma^2)$ as follows:

$$L(\underline{r_s}, \widetilde{S}, \sigma^2) = -\frac{PN}{2} \log(\sigma^2) - \frac{1}{\sigma^2} \sum_{k=0}^{\frac{N}{2}-1} ||\underline{g}(k)||^2 \tag{3.13}$$

and the corresponding ML estimates are:

$$\begin{aligned} \left(\widehat{\underline{r_s}}, \widehat{\widetilde{S}}, \widehat{\sigma^2}\right) &= \arg \max_{(\underline{r_s}, \widetilde{S}, \sigma^2)} \left\{ L(\underline{r_s}, \widetilde{S}, \sigma^2) \right\} \\ &= \arg \min_{(\underline{r_s}, \widetilde{S}, \sigma^2)} \left(\sum_{k=0}^{\frac{N}{2}-1} ||\underline{g}(k)||^2 \right) \\ &= \arg \min_{(\underline{r_s}, \widetilde{S}, \sigma^2)} \left(\sum_{k=0}^{\frac{N}{2}-1} \left[\underline{X}(k) - \widetilde{D}(k)\underline{S}(k)\right]^H \times \left[\underline{X}(k) - \widetilde{D}(k)\underline{S}(k)\right] \right) \end{aligned} \tag{3.14}$$

Thus, according to Eq. 3.14, we can write:

$$\widehat{\underline{S}} = \widetilde{D}(k)^{\dagger} \underline{X}(k) = \left(\widetilde{D}(k)^H \widetilde{D}(k)\right)^{-1} \widetilde{D}(k)^H \underline{X}(k) \tag{3.15}$$

and

$$\widehat{\underline{r_s}} = \arg\min_{\underline{r_s}} \sum_{k=0}^{\frac{N}{2}-1} ||\underline{X}(k) - \widetilde{D}(k)^{\dagger} \underline{X}(k)||^2 \tag{3.16}$$

3.3.2 Maximum Likelihood for Source Localization in the Presence of Nonuniform Noise Variances (SNWN)

In this subsection, we will introduce the nonuniform ML source localization formulation according to the recent literature [7, 8] for a more realistic SNWN model. Let $\underline{r_s}$, $\widetilde{S}$, and $\underline{q}$ be the parameters to be estimated for this case, where:

$$\underline{q} \stackrel{\text{def}}{=} [q_1, \ldots, q_P]^T \in \mathbb{R}^{P \times 1}$$

is the vector consisting of the diagonal elements in given by Eq. 3.8. The likelihood function of $(\underline{r_s}, \widetilde{S}, \underline{q})$ can be expressed as:

$$f\left(\underline{r_s}, \widetilde{S}, \underline{q}\right) \stackrel{\text{def}}{=} \frac{1}{\left(\pi^P \det(\widetilde{Q})\right)^{\frac{N}{2}}} \times \exp\left\{-\sum_{k=0}^{\frac{N}{2}-1} \underline{g}(k)^H \widetilde{Q}^{-1} \underline{g}(k)\right\} \tag{3.17}$$

Then, we have the following log-likelihood function $L(\underline{r_s}, \widetilde{S}, \underline{q})$ by taking the logarithm of Eq. 3.17 and neglecting all the constant terms

$$L(\underline{r_s}, \widetilde{S}, \underline{q}) = -\frac{N}{2} \sum_{p=1}^{P} \log(q_p) - \sum_{k=0}^{\frac{N}{2}-1} ||\underline{\dot{g}}(k)||^2 \tag{3.18}$$

where:

$$\underline{\dot{g}}(k) \stackrel{\text{def}}{=} \widetilde{Q}^{-\frac{1}{2}} \underline{g}(k) = \underline{\dot{X}}(k) - \dot{\widetilde{D}}(k)\underline{S}(k) \tag{3.19}$$

$$\underline{\dot{X}}(k) \stackrel{\text{def}}{=} \widetilde{Q}^{-\frac{1}{2}} \underline{X}(k) \tag{3.20}$$

$$\dot{\widetilde{D}}(k) \stackrel{\text{def}}{=} \widetilde{Q}^{-\frac{1}{2}} \widetilde{D}(k) \tag{3.21}$$

Consequently, we may obtain the ML estimates for $(\underline{r_s}, \widetilde{S}, \underline{q})$ as:

$$\left(\underline{\widehat{r_s}}, \widehat{\widetilde{S}}, \underline{\widehat{q}}\right) = \arg \max_{(\underline{r_s}, \widetilde{S}, \underline{q})} L(\underline{r_s}, \widetilde{S}, \underline{q}) \tag{3.22}$$

Similar to the derivation in Sect. 3.1, we can obtain the estimate of the pth element in $\underline{q}$ as:

$$\widehat{q}_p = \frac{2}{N} \sum_{k=0}^{\frac{N}{2}-1} |g_p(k)|^2 = \frac{2}{N} ||\underline{g_p}||^2 \tag{3.23}$$

where $g_p(k)$ denotes the pth element of the residual vector $\underline{g}(k)$ and

$$\underline{g_p} \stackrel{\text{def}}{=} \left[g_p(0), \ldots, g_p\left(\frac{N}{2} - 1\right)\right]^T \in \mathbb{C}^{\frac{N}{2} \times 1} \tag{3.24}$$

Substituting Eqs. 3.24 and 3.23 into Eq. 3.18, we can convert the log-likelihood function to a new version in terms of $\underline{r_s}$ and $\widetilde{S}$ and then get the ML estimators for $\underline{r_s}$ and $\widetilde{S}$ given by:

$$\left(\widehat{\underline{r_s}}, \widehat{\underline{\widetilde{S}}}\right) = \arg\max_{(\underline{r_s}, \underline{\widetilde{S}})} \left(-\sum_{p=1}^{P} \log ||\underline{g_p}||^2 \right) \tag{3.25}$$

and

$$\widehat{\underline{S}}(k) = \widetilde{\underline{\widetilde{D}}}(k)^{\dagger} \widetilde{\underline{\widetilde{X}}}(k) \tag{3.26}$$

Substituting Eqs. 3.25 and 3.26, we can obtain the ML estimates of $\underline{r_s}$ as:

$$\left(\widehat{\underline{r_s}}\right) = \arg\max_{(\underline{r_s})} \left(-\sum_{p=1}^{P} \log ||\underline{g_p}||^2 \right) \tag{3.27}$$

where $\underline{g_p}$ is defined by Eq. 3.24, and

$$\underline{g}(k) = \underline{X}(k) - \widetilde{D}(k) \widetilde{\underline{\widetilde{D}}}(k)^{\dagger} \widetilde{\underline{\widetilde{X}}}(k) \tag{3.28}$$

3.4 EM Widepand Source Localization Algorithm for Distinct Noise Variances

3.4.1 *Individual Likelihood Formulation for Source Localization*

The EM algorithm is a well-known iterative algorithm for the ML estimation. The complicated nonlinear optimization problem in Eqs. 3.22 and 3.27 can be simplified using the EM procedure incorporated with the *augmented (complete) data* corresponding to the individual incident source signals. First, we denote the received signal spectrum from the mth source to the pth sensor as $X_p^{(m)}(k)$ (for every $p = 1 \ldots P, m = 1 \ldots M$, and $k = 0 \ldots N-1$). Then, we define the augmented data as the following set:

$$\left\{ \underline{X}^{(m)}(k) \middle| m = 1 \ldots M \wedge k = 0 \ldots N-1 \right\}$$

where:

$$\underline{X}^{(m)}(k) \stackrel{\text{def}}{=} \left[X_1^{(m)}(k), \ldots, X_P^{(m)}(k) \right]^T \in \mathbb{C}^{P \times 1}$$

In addition, the relationship between the observed (incomplete) data $\underline{X}(k)$ and the unobserved latent (complete) data is established as:

$$\underline{X}(k) = \sum_{m=1}^{M} \underline{X}^{(m)}(k) \tag{3.29}$$

According to Eqs. 3.2, 3.5, 3.7, and 3.29, for a single source signal (the mth source), we have:

$$\underline{X}^{(m)}(k) \stackrel{\text{def}}{=} \underline{d}^{(m)}(k) S^{(m)}(k) + \underline{U}^{(m)}(k) \quad \forall k = 0 \ldots \frac{N}{2} - 1 \tag{3.30}$$

where $\underline{U}^{(m)}(k) \in \mathbb{C}^{P \times 1}$ is the complex-valued zero-mean uncorrelated Gaussian noise in the sole presence of the mth source.

According to Eqs. 3.22, 3.27, and 3.30, we have:

$$\left(\widehat{\underline{r_s}}^{(m)}, \widehat{\underline{S}}^{(m)}, \widehat{\underline{q}}^{(m)}\right) = \arg \max_{\left(\underline{r_s}^{(m)}, \underline{S}^{(m)}, \underline{q}^{(m)}\right)} L\left(\underline{r_s}^{(m)}, \underline{S}^{(m)}, \underline{q}^{(m)}\right) \quad \forall m = 1 \ldots M \tag{3.31}$$

where:

$$\underline{S}^{(m)} \stackrel{\text{def}}{=} \left[S^{(m)}(0), \ldots, S^{(m)}(\frac{N}{2} - 1)\right]^T \in \mathbb{C}^{\frac{N}{2} \times 1}$$

and

$$\underline{q}^{(m)} \stackrel{\text{def}}{=} \left[q_1^{(m)}, \ldots, q_P^{(m)}\right]^T \in \mathbb{C}^{P \times 1}$$

is the vector consisting of the diagonal elements in $\widetilde{Q}^{(m)}$ which is defined in the following form:

$$\widetilde{Q}^{(m)} \stackrel{\text{def}}{=} E\left[\underline{U}^m(k)\left(\underline{U}^{(m)}(k)\right)^H\right] \in \mathbb{C}^{P \times P}$$

Now, let $\underline{\dot{d}}^{(m)}(k)$ denote the following expression:

$$\underline{\dot{d}}^{(m)}(k) \stackrel{\text{def}}{=} \left(\widetilde{Q}^{(m)}\right)^{\frac{1}{2}} \underline{d}^{(m)}(k) \tag{3.32}$$

and

$$\underline{\dot{X}}^{(m)}(k) \stackrel{\text{def}}{=} \left(\widetilde{Q}^{(m)}\right)^{\frac{1}{2}} \underline{X}^{(m)}(k) \tag{3.33}$$

According to Eq. 3.24, we denote the pth element of the particular residual vector $\underline{g}^{(m)}(k)$ as $g_p^{(m)}(k)$ when only source m is present, where:

$$\underline{g}^{(m)}(k) = \underline{X}^{(m)}(k) - \underline{\dot{d}}^{(m)}(k) \underline{\dot{d}}^{(m)}(k)^{\dagger} \underline{\dot{X}}^{(m)}(k) \tag{3.34}$$

Similar to the derivation in Sect. 3.2, Eq. 3.31 yields:

$$\widehat{q_p}^{(m)} = \frac{2}{N} \sum_{k=0}^{\frac{N}{2}-1} \left| \left[g_p^{(m)}(k) \right] \right|^2 = \frac{2}{N} ||\underline{g_p}^{(m)}||^2 \tag{3.35}$$

where:

$$\underline{g_p}^{(m)} \stackrel{\text{def}}{=} \left[g_p^{(m)}(0), \ldots, g_p^{(m)}\left(\frac{N}{2}-1\right) \right]^T \in \mathbb{C}^{\frac{N}{2}\times 1} \tag{3.36}$$

Consequently, the ML estimates $\underline{\widehat{r_s}}^{(m)}$ is given by

$$\left(\underline{\widehat{r_s}}^{(m)}\right) = \arg \max_{(\underline{r_s}^{(m)})} \left(- \sum_{p=1}^{P} \log \left(||\underline{g_p}^{(m)}||^2 \right) \right) \tag{3.37}$$

According to Eq. 3.37, the source localization problem can be formulated as the independent maximization subproblems with respect to the individual likelihood functions.

3.4.2 New Expectation–Maximization Algorithm for Source Localization

In contrast to other existing algorithms for the source localization using the sensor signals in the presence of noises with identical variance [1, 3, 17, 18], we present a new EM algorithm here to solve the realistic source localization problem for sensor signals in the presence of noises with different variances, which has been tackled by [7] recently. Nevertheless, our proposed EM algorithm can be demonstrated to be more robust than the method proposed by [7].

The details of our proposed EM algorithm are introduced as follows (since our proposed algorithm can be decoupled across different sources in each iteration, we only need to address the steps for the source m and it can be run for other sources as well in parallel).

Initialization: Randomly initialize $[\underline{\widehat{r_s}}^{(m)}]^{[0]}$. Set the initial values for the entries in $[\underline{\widehat{q}}^{(m)}]^{[0]}$ and $[\underline{\widehat{q}}]^{[0]}$ as:

$$\left[\underline{\widehat{q}}^{(m)}\right]^{[0]} = \frac{1}{M} \times [11\ldots1]^T \in \mathbb{R}^{P\times 1} \tag{3.38}$$

$$\left[\underline{\widehat{q}}\right]^{[0]} = [11\ldots1]^T \in \mathbb{R}^{P\times 1} \tag{3.39}$$

respectively.

Input (Given) Parameters at Iteration i: $[\underline{\widehat{q}}^{(m)}]^{[i-1]}$, $[\underline{\widehat{r_s}}^{(m)}]^{[i-1]}$.
Output Variables at Iteration i: $[\underline{\widehat{q}}^{(m)}]^{[i]}$, $[\underline{\widehat{r_s}}^{(m)}]^{[i]}$.
Given the input parameters, the EM algorithm for the ith iteration is stated next.
Expectation Step (E-Step): Calculate

$$\widehat{\widehat{Q}}^{(m)} = \text{diag}\left\{\left[\underline{\widehat{q}}^{(m)}\right]^{[i-1]}\right\} \tag{3.40}$$

where diag{} converts the vector inside the associated braces into a diagonal matrix containing the vector's entries as the diagonal elements in the same order. Compute

$$\widetilde{Q} = \sum_{m=1}^{M} \widehat{\widehat{Q}}^{(m)} \tag{3.41}$$

and

$$\alpha = \frac{\left[\text{trace}\left(\widehat{\widehat{Q}}^{(m)}\right)\right]^2}{\left[\text{trace}\left(\widetilde{\widetilde{Q}}\right)\right]^2} \tag{3.42}$$

Calculate

$$t_p^{(m)} = F_s \frac{\left\|\left[\underline{\widehat{r_s}}^{(m)}\right]^{[i-1]} - \underline{r_p}\right\|}{v} \tag{3.43}$$

According to Eqs. 3.4–3.6, 3.43, and $a_p^{(m)} = 1,\ \forall p$, based on [7], determine $\underline{d}^{(m)}(k)$ and $\widetilde{D}(k)$. Next, follow Eqs. 3.20, 3.21, and 3.26 to determine $\widehat{\underline{S}}(k)$ and $\widehat{\overline{S}}^{(m)}(k)$, $k = 0, \ldots, \frac{N}{2} - 1$, where $\widehat{S}^{(m)}(k)$ is the mth element of $\widehat{\underline{S}}(k)$. Then, determine

$$\begin{aligned}\widehat{\underline{X}}^{(m)} &= E\left[\widehat{\underline{X}}^{(m)}(k)\middle|\underline{X}(k)\right]\\ &= \underline{d}^{(m)}(k)\widehat{S}^{(m)}(k) + \alpha\left(\underline{X}(k) - \widetilde{D}(k)\widehat{\underline{S}}(k)\right) \quad \forall k = 0, \ldots, \frac{N}{2} - 1\end{aligned} \tag{3.44}$$

Maximization Step (M-Step): Now, let

$$t_p^{(m)} = F_s \frac{\left\|\underline{r_s}^{(m)} - \underline{r_p}\right\|}{v} \tag{3.45}$$

$$\underline{\dot{d}}^{(m)}(k) = \left(\widehat{\widehat{Q}}^{(m)}\right)^{(-\frac{1}{2})} \underline{d}^{(m)}(k) \quad \forall k = 0, \ldots, \frac{N}{2} - 1 \tag{3.46}$$

which also involves the variable coordinate $\underline{r_s}^{(m)}$. According to the result from Eq. 3.44, calculate

$$\widehat{\underline{\tilde{X}}}^{(m)}(k) = \left(\widehat{\underline{Q}}^{(m)}\right)^{\left(-\frac{1}{2}\right)} \widehat{\underline{X}}^{(m)}(k) \quad \forall k = 0, \ldots, \frac{N}{2} - 1 \tag{3.47}$$

Then, construct

$$\widehat{\underline{g}}^{(m)}(k) = \widehat{\underline{\tilde{X}}}^{(m)}(k) - \underline{d}^{(m)}(k)\hat{\underline{d}}^{(m)}(k)^{\dagger}\widehat{\underline{\tilde{X}}}^{(m)}(k) \quad \forall k = 0, \ldots, \frac{N}{2} - 1 \tag{3.48}$$

which involves the variable coordinate $\underline{r_s}^{(m)}$ as well. Denote the pth element of $\widehat{\underline{g}}^{(m)}(k)$ as $\widehat{g}_p^{(m)}$. Facilitate

$$\widehat{\underline{g_p}}^{(m)} = \left[\widehat{g}_p^{(m)}(0), \ldots, \widehat{g}_p^{(m)}\left(\frac{N}{2} - 1\right)\right]^T \tag{3.49}$$

which involves the variable coordinate $\underline{r_s}^{(m)} \in \mathbb{R}^{2\times1}$. Carry out

$$\left[\widehat{\underline{r_s}}^{(m)}\right]^{[i]} = \arg\min_{\underline{r_s}^{(m)}} \sum_{p=1}^{P} \log\left(||\widehat{\underline{g_p}}^{(m)}||^2\right) \tag{3.50}$$

Besides, calculate $t_p^{(m)}$ using Eq. 3.43. Let $a_p^{(m)} = 1, \ \forall p$. Enumerate the parameters given by Eqs. 3.5, 3.6, 3.32, 3.44, 3.47–3.49 in this sequential order. Then, calculate

$$\left[\widehat{q}_p^{(m)}\right]^{[i]} = \frac{2}{N}||\widehat{\underline{g_p}}^{(m)}||^2 \quad \forall p = 1, \ldots P \tag{3.51}$$

Thus, obtain

$$\left[\widehat{\underline{q}}^{(m)}\right]^{[i]} = \left[\left[\widehat{q}_1^{(m)}\right]^{[i]}, \ldots, \left[\widehat{q}_P^{(m)}\right]^{[i]}\right]^T \in \mathbb{R}^{P\times1} \tag{3.52}$$

The above algorithm facilitates a recursive solution to multiple wideband source localization.

3.5 Robustness Analysis for Source Localization Algorithms

3.5.1 Localization Algorithms

CRLB is often used to characterize the robustness of the estimation methods. In this section, by extending the CRLB presented in [7] for the simple DOA estimation problem, we derive the CRLB for the source localization problem to benchmark our EM method and the SC-ML/AC-ML schemes as

$$\frac{1}{\text{CRLB}} = 2\Re\left\{\sum_{k=0}^{\frac{N}{2}-1}\left\{\left[\tilde{\hat{G}}(k)^H \widetilde{P}^{\perp}_{\tilde{\tilde{D}}(k)} \tilde{\hat{G}}(k)\right] \odot \widetilde{R}_s(k)^T\right\}\right\} \tag{3.53}$$

where

$$\tilde{\hat{G}}(k) \stackrel{\text{def}}{=} \widetilde{Q}^{-\frac{1}{2}} \widetilde{G}(k) \tag{3.54}$$

$$\widetilde{G}(k) \stackrel{\text{def}}{=} \left[\frac{\partial}{\partial \underline{r_s}^{(1)}} \underline{d}^{(1)}(k), \ldots, \frac{\partial}{\partial \underline{r_s}^{(M)}} \underline{d}^{(M)}(k)\right] \tag{3.55}$$

$$\widetilde{P}^{\perp}_{\tilde{\tilde{D}}(k)} \stackrel{\text{def}}{=} \widetilde{I} - \tilde{\tilde{D}}(k)\tilde{\tilde{D}}(k)^{\dagger} \tag{3.56}$$

$$\widetilde{R}_s(k) \stackrel{\text{def}}{=} \underline{S}(k)\underline{S}(k)^H \tag{3.57}$$

Note that $\widetilde{Q}$, $\underline{d}^{(m)}(k)$, and $\underline{S}(k)$ are given by Eqs. 3.4, 3.5, 3.7, 3.8, 3.21. We can rewrite 3.55 as

$$\widetilde{G}(k) = \frac{\partial \widetilde{D}(k)}{\partial \underline{r_s}^T} = -jF_s k \frac{2\pi}{Nv} \times \widetilde{F} \tag{3.58}$$

where

Note that $\underline{r_s}^{(m)} \stackrel{\text{def}}{=} [\chi_s^{(m)}, y_s^{(m)}]^T$ and $\underline{r_p} \stackrel{\text{def}}{=} [\chi_p, y_p]^T$.

3.6 Simulation

The comparison is made among our newly proposed EM-based multiple wideband source localization scheme, the SC-ML method and the AC-ML method here. The sampling frequency is 100 kHz. The propagation speed is 345 meters/s. The data are simulated for a circularly shaped array of five sensors using the recorded acoustic data acquired from [1] as shown in Fig. 3.1 (squares denote the sensor locations and circles denote the actual source locations). The sample size is $L = 200$ and the DFT size is $N = 256$. Throughout the simulation, the minimization in our EM method characterized by (50) is performed by Nelder–Mead direct search [3], while the optimization steps in both SC-ML and AC-ML methods are performed using the AM algorithm, which would lead to better performance than Nelder–Mead direct search in these two schemes [3, 7]. Moreover, the additive noises in all experiments are randomly generated by a Gaussian process using the computer, and the signal-to-noise ratio (SNR) is defined according to [7, 8].

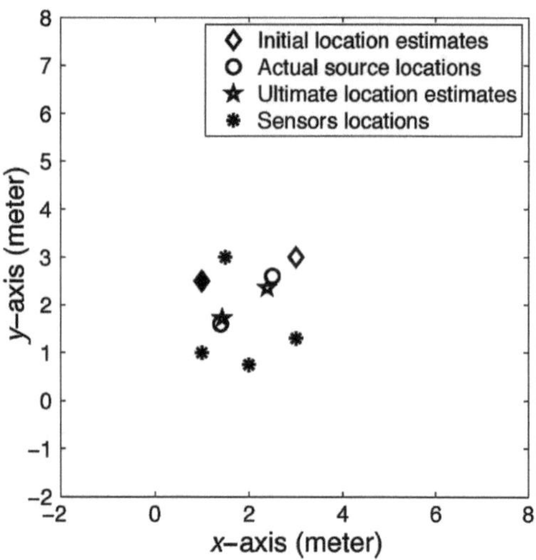

Fig. 3.1 The localization of two wideband (acoustic) sources in the near field corrupted by the noises with nonuniform variances (SNR is 10 dB). The initial location estimates and the ultimate location estimates resulted from the EM algorithm (three iterations are taken) are also demonstrated

3.6.1 A Localization Layout Example

Then we investigate the performance of the EM algorithm for estimating the two source locations in the presence of sensor noises with nonuniform variances and compare with the SC-ML and AC-ML algorithms. The noise processes across different sensors have the covariance matrix as $\widetilde{Q} = \sigma^2 \mathrm{diag}\{2, 3, 1, 5, 9\}$. One hundred Monte Carlo experiments are carried out using our EM method with randomly initialized source locations for a particular signal-to-noise ratio (SNR = 10 dB). The localization result from a certain experiment is depicted in Fig. 3.1, where the ultimate locations are achieved after three iterations of EM algorithm. We default the number of EM iterations as three in all Monte Carlo experiments.

3.6.2 Root-Mean-Square Errors and Computational Complexities for Source Localization

For each SNR value ranging from 0 to 40 dB, we fix the initial source location estimates as depicted in Fig. 3.1 and carry out a hundred Monte Carlo experiments to obtain the average localization accuracy in terms of the RMS error in meters. The three corresponding RMS error curves to the three aforementioned schemes are depicted in Fig. 3.2a. Then, we vary the initial location estimates around the circular areas with a one-meter diameter with respect to the two initial source location estimates depicted in Fig. 3.1 and redo 100 Monte Carlo experiments similar to the setup generating Fig. 3.2a. The results are depicted in Fig. 3.2b. It is obvious that the accuracies of all three methods degrade from Fig. 3.2a and b since the initial

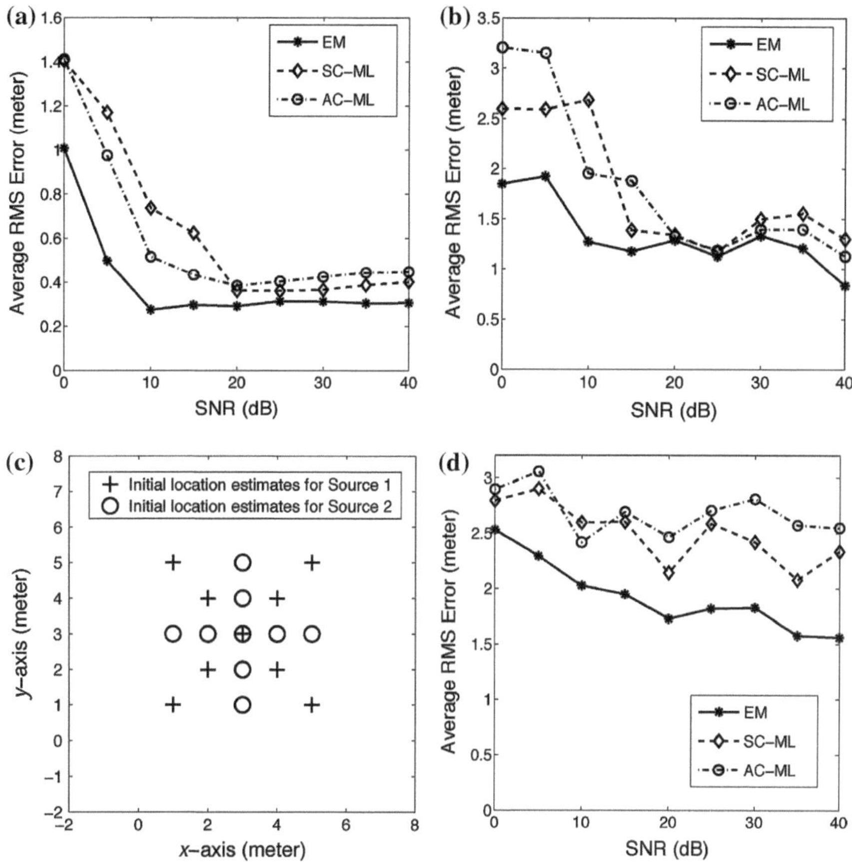

Fig. 3.2 Average RMS localization errors versus SNR and 18 different initial source location estimates. **a** Average RMS localization errors versus SNR for the sources corrupted by the noises with nonuniform variances. The initial location estimates are plotted in Fig. 3.1. **b** Average RMS localization errors versus SNR for the sources corrupted by the noises with nonuniform variances. The initial source location estimates here are randomly chosen within the areas which are one meter around the initial location estimates used in Fig. 3.1. **c** The 18 different initial source location estimates. **d** Average RMS localization errors versus SNR for the sources corrupted by the noises with nonuniform variances. The initial source location estimates are plotted in Fig. 3.2c

conditions change. To further study this effect, we spread the initial location estimates over a broader area as depicted in Fig. 3.2c and redo 100 Monte Carlo experiments similar to Fig. 3.2b. The average RMS error curves are demonstrated in Fig. 3.2d. Next, we would like to investigate the performances of the three aforementioned localization methods for the sensor noises with identical variances (SWN). Thus, we choose the sensor noise covariance matrix as $\widetilde{Q} = \sigma^2\text{diag}\{1, 1, 1, 1, 1\}$ now. With this new noise covariance matrix, we redo the Monte Carlo experiments similar to those generating Fig. 3.2a, b, and d. The corresponding results are plotted in Fig. 3.3a, b, and c, respectively. According to these two sets of experiments, our proposed EM

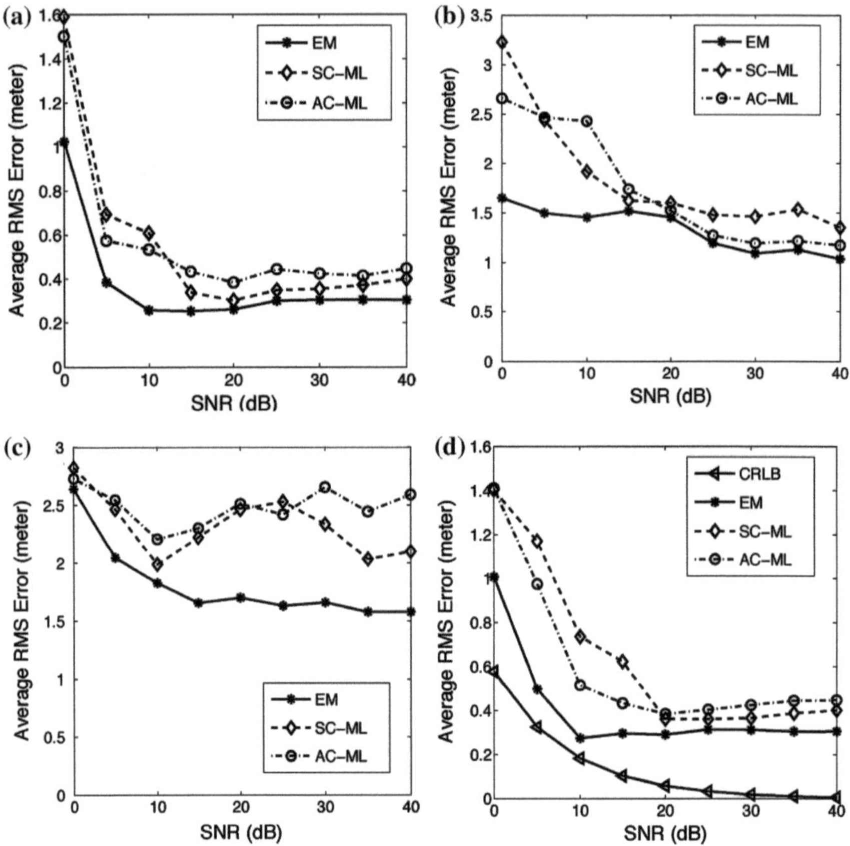

Fig. 3.3 Average RMS localization errors versus SNR. **a** Average RMS localization errors versus SNR for the sources corrupted by the noises with identical variances. The initial source location estimates are plotted in Fig. 3.1. **b** Average RMS localization errors versus SNR for the sources corrupted by the noises with identical variances. The initial source location estimates are randomly drawn from the areas which are one meter around the initial source location estimates in Fig. 3.1. **c** Average RMS localization errors versus SNR for the sources corrupted by the noises with identical variances. The initial source location estimates are plotted in Fig. 3.2c. **d** CRLBs and simulated RMS localization errors (actual variances) versus different SNR values for the three schemes in comparison

algorithm greatly outperforms both SC-ML and AC-ML methods in all conditions. In addition, the accuracies of all three methods degrade due to the changes in the initial conditions for the SWN scenario as well. Besides, the performances of all these three schemes for the SWN case are not much different from those for the SNWN case, since the SWN model is a particular case of the SNWN model.

3.6.3 Robustness Analysis of Source Localization

We fix the initial source location estimates as those generating Fig. 3.1 and carry out a hundred Monte Carlo experiments again. The corresponding CRLBs for our EM method, the SC-ML (or AC-ML) method, are depicted in Fig. 3.3d. We also depict the average RMS error curves in the same figure. According to Fig. 3.3d, we discover that the RMS errors resulted from our EM algorithm are much closer to the CRLBs than the SC-ML and AC-ML methods. Note that all the three source localization schemes in comparison are quite sensitive to the initial condition. This still remains as a very challenging problem for the wideband source localization. Note that our experimental results illustrated in this chapter can be generalized for other conditions. It means that if we change the source locations and use all the three algorithms subject to the same initial conditions, the experimental results under every different condition specified in Sects. 3.6.1–3.6.3 will be very similar to Figs. 3.2 and 3.3.

3.7 Summary and Outlook

In this chapter, we propose a novel EM-based multiple wideband source localization scheme in the presence of nonuniform noise variances. For our EM method and the conventional SC-ML and AC-ML methods, the performance is rather sensitive to the initial source location estimates. Our proposed EM algorithm can lead to an outstanding localization performance given a reasonably good initial condition. Moreover, our proposed EM algorithm can always outperform the conventional SC-ML and AC-ML methods when the initial source location estimates are randomly chosen. The Monte Carlo simulation results demonstrate the superiority of our proposed EM method. To provide the robustness analysis for the source localization algorithms, we present the CRLB associated with these three schemes. The CRLB analysis demonstrates that our proposed EM algorithm is much closer to the achievable minimum variance than the two other methods in all SNR conditions. In addition, according to our complexity analysis, the complexity measure for our proposed algorithm is of $O(M^2)$ which is much less than those for the SC-ML and AC-ML methods [both with a complexity measure of $O(M^3)$.

References

1. K. Yan, H.-C. Wu, S.S. Iyengar, Robustness analysis of source localization using Gaussianity measure, in *Proceedings of IEEE Global Telecommunication Conference*, Nov 2008, pp. 1–5
2. H. Krim, M. Viberg, Two decades of array signal processing research: The parametric approach. IEEE Signal Process. Mag. **13**(4), 67–94 (1996)
3. J.C. Chen, R.E. Hudson, K. Yao, Maximum-likelihood source localization and unknown sensor location estimation for wideband signals in the near-field. IEEE Trans. Signal Process. **50**(8), 1843–1854 (2002)

4. J.C. Chen, R.E. Hudson, K. Yao, Source localization of a wideband source using a randomly distributed beamforming sensor array, in *Proceedings of the International Society of Information Fusion*, 2001, pp. TuC1:11–TuC1:18
5. P. Stoica, A. Nehorai, Music, maximum likelihood and Cramer-Rao bound. IEEE Trans. Acoust. Speech Signal Process. **37**(5), 720–741 (1989)
6. P. Stoica, A. Nehorai, Performance study of conditional and unconditional direction-of-arrival estimation. IEEE Trans. Acoust. Speech Signal Process. **38**(10), 1783–1795 (1990)
7. C.E. Chen, F. Lorenzelli, R.E. Hudson, K. Yao, Maximum likelihood DOA estimation of multiple wideband sources in the presence of nonuniform sensor noise. EURASIP J. Adv. Signal Process.
8. M. Pesavento, A.B. Gershman, Maximum-likelihood direction- of-arrival estimation in the presence of unknown nonuniform noise. IEEE Trans. Signal Process. **49**(7), 1310–1324 (2001)
9. A.P. Dempster, N.M. Laird, D.B. Rubin, Maximum likelihood from incomplete data via the EM algorithm. J. Royal Statist. Soc. **39**(1), 1–38 (1977)
10. M. Feder, E. Weinstein, Multipath and multiple source array processing via the EM algorithm. Proc. IEEE Int. Conf. Acoust. Speech Signal Process. **11**, 2503–2506 (1986)
11. M. Feder, E. Weinstein, Parameter estimation of superimposed signals using the EM algorithm. IEEE Trans. Acoust. Speech Signal Process. **36**(4), 477–489 (1988)
12. E. Weinstein, V. Oppenheim, M. Feder, Iterative and sequential algorithms for multisensor signal enhancement. IEEE Trans. Signal Process. **42**(4), 846–859 (1994)
13. P.J. Chung, J.F. Bohme, Recursive EM and SAGE algorithms, in *Proceedings of IEEE Workshop Statistical Signal Processing*, 2001, pp. 540–543
14. P.J. Chung, J.F. Bohme, A.O. Hero, Tracking of multiple moving sources using recursive EM algorithm. EURASIP J. Appl. Signal Process. **1**(1), 50–60 (2005)
15. L. Frenkel, M. Feder, Recursive expectation-maximization (EM) algorithms for time-varying parameters with applications to multiple target tracking. IEEE Trans. Signal Process. **47**(2), 306–320 (1999)
16. M.I. Miller, D.R. Fuhrmann, Maximum-likelihood narrow-band direction finding and the EM algorithm. IEEE Trans. Acoust. Speech Signal Process. **38**(9), 1560–1577 (1990)
17. K.K. Mada, H.-C. Wu, EM algorithm for multiple wideband source, in *Proceedings of IEEE Global Telecommunication Conference*, 2006, pp. 1–5
18. K.K. Mada, H.-C. Wu, S.S. Iyengar, Efficient and robust EM algorithm for multiple wideband source localization. IEEE Trans. Veh. Technol. **58**(6), 3071–3075 (2009)
19. L. Lu, H.C. Wu, K. Yan, S.S. Iyengar, Robust expectation-maximization algorithm for multiple wideband acoustic source localization in the presence of nonuniform noise variances. IEEE Sens. J. **11**(3), 536–544 (2011)

Part II
Coverage Problems

Chapter 4
Coordinate-Free Coverage in Sensor Networks via Homology

We introduce tools from computational homology to verify coverage in an idealized sensor network. Our methods are unique in that, while they are coordinate-free and assume no localization or orientation capabilities for the nodes, there are also no probabilistic assumptions. The key ingredient is the theory of homology from algebraic topology. We demonstrate the robustness of these tools by adapting them to a variety of settings, including static planar coverage, 3-D barrier coverage, and time-dependent sweeping coverage. We also give results on hole repair, error tolerance, optimal coverage, and variable radii. An overview of implementation is given.

4.1 Introduction

Sensor networks are an increasingly essential and pervasive feature of modern computation and automation [15]. Within this large topic of active and rapidly developing research, *coverage problems* are common. Such problems, involving gaps or holes in sensor networks, appear in a variety of settings relevant to robotics and networks: environmental sensing, communication and broadcasting, robot beacon navigation, surveillance, security, and warfare are common application domains. A specific example is as follows. Given a collection of nodes $\mathcal{X}$ in a bounded domain $\mathcal{D}$ of the plane, assume that each node can sense, broadcast to, or otherwise cover a region of fixed *coverage radius* about the node. The most basic form of coverage problem is the simple query: given the nodes, does the collection of coverage disks at $\mathcal{X}$ cover the domain $\mathcal{D}$?

We provide a sufficiency criterion for coverage. We do not answer the problem of how the nodes should be placed in order to maximize coverage—nodes are assumed to be distributed a priori, yet not according to some fixed protocol. In particular,

---

This chapter has been reprinted with permission from "Coordinate-Free Coverage in Sensor Networks with Controlled Boundaries via Homology," V. De Silva, R. Ghrist, The International Journal of Robotics Research, January 2006.

S. S. Iyengar et al., *Mathematical Theories of Distributed Sensor Networks*,
DOI: 10.1007/978-1-4419-8420-3_4,

there are no assumptions about random distributions or densities. The coverage criterion we introduce is both computable and, at this time, centralized. We do not here demonstrate how to reduce the homological criteria of this chapter to a distributed computation.

4.1.1 Preliminary Assumptions

We assume a complete absence of localization capabilities. *Nodes can determine neither distance nor direction.* Only connectivity data between nodes are used. The only strong assumption we make is on the *fence nodes* setup along the boundary of the domain. This strong degree of control along the boundary is not strictly required (see Sect. 4.6 and also [11]), but it simplifies the statements and proofs of theorems dramatically.

Assumption A1 Nodes $\mathcal{X}$ broadcast their unique ID numbers. Each node can detect the identity of any node within *broadcast radius* r_b.

Assumption A2 Nodes have radially symmetric covering domains of *cover radius* $r_c \geq \frac{r_b}{\sqrt{3}}$.

Assumption A3 Nodes $\mathcal{X}$ lie in a compact connected domain $\mathcal{D} \subseteq \mathbb{R}^2$ whose boundary $\partial\mathcal{D}$ is connected and piecewise-linear with vertices marked *fence nodes* $\mathcal{X}_f$.

Assumption A4 Each fence node $v \in \mathcal{X}_f$ knows the identities of its neighbors on $\partial\mathcal{D}$ and these neighbors both lie within distance r_b of v.

To summarize, the sensor data for each node consists of a list of node ID numbers within signal detection range, as well as a binary flag denoting whether or not it is a marked fence node.

4.1.2 Results

We claim that, surprisingly, such coarse coordinate-free data are sufficient to rigorously verify coverage in many instances. One constructs the *communication graph* whose vertices are the nodes of the network and whose edges represent signal detection connectivity (at radius r_b). From this graph, we build the *Rips complex* $\mathcal{R}$: the largest simplicial complex with the corresponding graph as its one-dimensional skeleton. By Assumption **A4**, the boundary $\partial\mathcal{D}$ can be represented as a one-dimensional fence cycle $\mathcal{F} \subseteq \mathcal{R}$, which is canonically identified with $\partial\mathcal{D}$.

Our results are all based on a certain algebraic-topological invariant of these simplicial complexes—homology. The following is the principal criterion for coverage we derive in this chapter:

Theorem 4.1 *The union of the radius r_c disks contains $\mathcal{D}$ if there is a nontrivial element of the relative homology $H_2(\mathcal{R}, \mathcal{F})$ whose boundary is non-vanishing.*

See Theorem 3.3 for details. The casual reader is advised to think of this homology $H_2(\mathcal{R}, \mathcal{F})$ as a vector space which is computed from the network according to some algorithm. The criterion of the Theorem 4.1 is that, first, this vector space has dimension greater than zero, and second, one can find a "good" basis element.

In Sects. 4.4–4.11, we provide several extensions of this result. These include the following:

1. Criteria for performing "hole repair" in systems for which the coverage criterion fails;
2. Criteria for localized coverage in an unbounded network resulting from querying a cycle in the communication graph;
3. Criteria for coverage in domains with multiple boundary components;
4. A homological approach to identifying redundant nodes in a cover;
5. Coverage criteria for systems with varying communication and coverage radii;
6. Coverage criteria for systems with communication errors and faulty nodes;
7. Barrier coverage for 3-D systems in a tunnel-like domain;
8. Pursuit-evasion criteria for time-dependent systems.

Comments on implementation and simulations appear in Sect. 4.12, followed by a discussion.

4.1.3 Related Work

There is a large literature on the subject of static or "blanket" coverage; see, e.g., [4, 17, 29] and references therein. In addition, there are variants on these problems involving "barrier" coverage to separate regions. Dynamic or "sweeping" coverage [9] is a common and challenging task with applications ranging from security to housekeeping.

There are two primary approaches to static coverage problems in the literature. The first uses computational geometry tools applied to exact node coordinates. This typically involves computational geometry [23] and Delaunay triangulations of the domain [27, 29, 38]. Such approaches are very rigid with regard to inputs: one must know exact node coordinates and one must know the geometry of the domain precisely to determine the Delaunay complex.

To alleviate the former requirement, many authors have turned to probabilistic tools. For example, in [25], the author assumes a randomly and uniformly distributed collection of nodes in a domain with a fixed geometry and proves expected area coverage. Other approaches [22, 26, 28, 37] give probabilistic or percolation results about coverage and network integrity for randomly distributed nodes. The drawback of these methods is the need for a uniform distribution of nodes.

More recently, the robotics community has explored how networked sensors and robots can interact and augment each other: see, e.g., [3, 5, 8, 15] and references therein. There are several new approaches to networks without localization that come from researchers in ad hoc wireless networks that are not unrelated to coverage questions. One example is the routing algorithm of [34], which generally works in practice but is a heuristic method involving heat-flow relaxation. The papers [6, 18, 31, 35] give methods for localizing an entire network if localization of a certain portion is known. More recent work of Fekete et al. [16] grows and merges cycles in a distributed manner to "fill up" a sufficiently well-sampled network to determine boundaries in a coordinate-free network. This is one example of the work in computational geometry concerning *unit disk graphs*.

The mathematical tools we introduce for coverage problems—homology theory—date roughly from the 1930s. The use of homology as an effective tool in scientific computation is more recent: see, e.g., the textbook of [24] and its references. Homology has recently been used is several applied contexts, from point cloud shape representation and high-dimensional data analysis [10, 39], vision [1], applied differential equations [24, 30], and hybrid controls [2].

4.2 The Rips Complex

Given a collection of nodes $\mathcal{X}$ in a domain, we wish to determine the global properties of $\mathcal{U}$, the union of coverage domains centered at these nodes. However, we are constrained to use only communication connectivity data between nodes. Instead of restricting attention to the graph of pairwise-connectivity data, we complete it to a higher-dimensional complex. This type of simplicial complex was introduced by Vietoris in the early history of homology theory [36] and has more recently been reinterpreted by Rips [20] and used extensively in geometric group theory.

Definition 4.1 Given a set of points $X = \{x_\alpha\}$ in a metric space and a fixed $\epsilon > 0$, the Rips complex of $\mathcal{X}$, $\mathcal{R}_\epsilon(\mathcal{X})$, is the abstract simplicial complex whose k-simplices correspond to unordered $(k+1)$-tuples of points in $\mathcal{X}$, which are pairwise within distance ϵ of each other.

Our goal is to compare the topology of the Rips complex $\mathcal{R} = \mathcal{R}_{r_b}(\mathcal{X})$ to the union of covering disks $\mathcal{U} = \mathcal{U}_{r_c}(\mathcal{X})$. The cover $\mathcal{U}$ is necessarily a subset of $\mathbb{R}^2$; the Rips complex, in contrast, may have any dimension, depending on clustering of nodes. It is best to visualize $\mathcal{R}$ as a high-dimensional space that "floats" above the Euclidean plane: cf. Fig. 4.1. This chapter asserts that topological features of $\mathcal{R}$ suffice to conclude geometric properties of $\mathcal{U}$.

The following lemma demonstrates that the choice of bound for rc in A2 is the appropriate one.

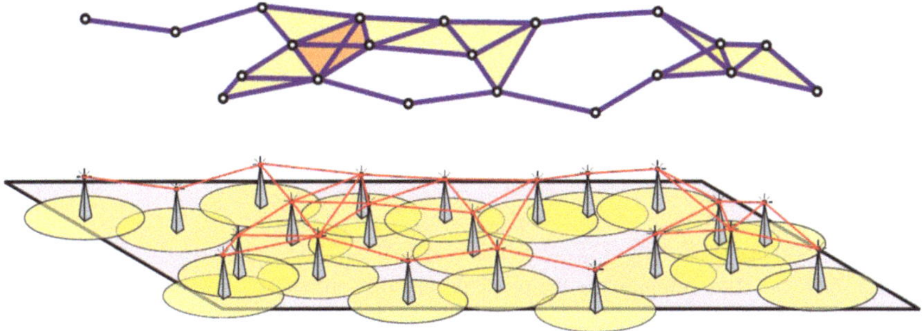

Fig. 4.1 A collection of sensor nodes generates a cover in the workspace (*bottom*). The Rips complex of the network is an abstract simplicial complex that has no localization or coordinate data (*top*). In the example illustrated, the Rips complex encodes the communication network as one-closed 3-simplex, eleven-closed 2-simplices, and seven-closed 1-simplices connected as shown. The "holes" in this Rips complex reflect the holes in the sensor cover, below

Lemma 4.1 *The convex hull of any collection of nodes in $\mathcal{D}$ which form a simplex of $\mathcal{R}$ lies within $\mathcal{U}$.*

Proof Any collection of circular disks that meet at a common point x necessarily covers the convex hull of x and the centers of the disks. So, it suffices to show that the balls of radius r_c intersect. It also suffices to prove this for a 2-simplex of $\mathcal{R}$ thanks to Helly's theorem [14], which implies that a collection of $k \geq 4$ convex sets in $\mathbb{R}^2$ has a nonempty common intersection provided only that the same is true for each subset of size 3.

Therefore, consider a triple of points $\{x_i\}_1^3$ which span a triangle with side lengths at most r_b. We must show that the three disks of radius r_c centered on $\{x_i\}_1^3$ meet at a common point. If the triangle is obtuse (or right-angled), then the midpoint of the longest side is common to all three disks; hence, $r_c \geq r_b/2$ suffices. If the triangle is acute, then the largest angle, say θ_1 at vertex x_1, satisfies $\pi/3 \leq \theta_1 \leq \pi/2$ and so $\sin(\theta_1) \geq \sqrt{3}/2$. We can compute the circumradius R of the triangle as:

$$R = \frac{||x_2 - x_3||}{2\sin(\theta_1)}$$

and hence, we deduce $R \leq r_b/\sqrt{3} \leq r_c$. Thus, in this case, the three disks meet at the circumcenter. □

Remark 4.1 The ratio $r_c \geq r_b/\sqrt{3}$ is optimal: consider an equilateral triangle of side length r_b.

Unfortunately, the radius-r_b Rips complex of a set of nodes in $\mathbb{R}^2$ does not always capture the topology of the union of radius-r_c balls centered on these nodes. Figure 4.2 gives examples for which the Rips complex fails to capture the topology of the cover.

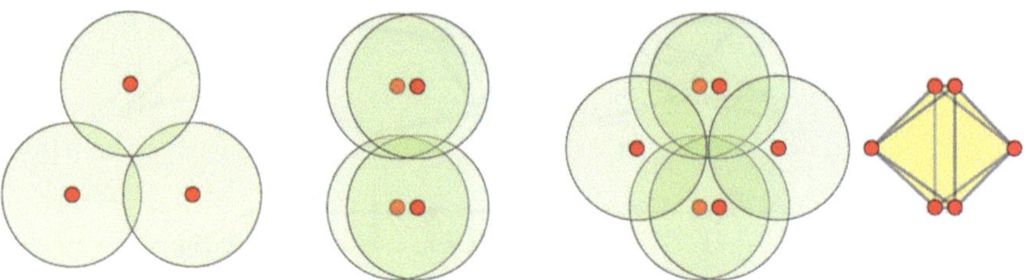

Fig. 4.2 *Left* The Rips complex has the property that all 2-simplices determine triangles in the domain which lie within the radius r_c cover. However, the Rips complex does not capture the topology of the cover. A contractible union of r_c balls can have Rips complex with nontrivial homology in dimension one (*center* in which $\mathcal{R}$ is a quadrilateral), two (*right* in which $\mathcal{R}$ is the boundary of a solid octahedron), or higher

4.3 A Homological Criterion for Coverage

We use the homology of $\mathcal{R}$ relative to $\mathcal{F}$ to obtain a coverage criterion.

The intuition behind the coverage criterion is very straightforward. Based on the communication graph alone, it is difficult to "see" potential holes in coverage. However, upon completing the graph to the Rips complex $\mathcal{R}$, large holes in coverage would seem to be present in the abstract complex: see Fig. 4.3. One might guess that showing there are no such holes in $\mathcal{R}$ implies coverage. This condition would be translated into algebraic-topological terms as $H_1(\mathcal{R}) = 0$, or, that any cycle in the communication graph can be realized as the boundary of a surface built from 2-simplices of $\mathcal{R}$, each of which indicates a coverage region thanks to Lemma 4.1.

We use a slightly different criterion than $H_1(\mathcal{R}) = 0$: one which is more robust to extensions and which yields stronger information about the actual cover. The fence cycle $\mathcal{F}$ is canonically identified with the boundary $\partial\mathcal{D}$. If this cycle is *null-homologous*—that is, if $[\mathcal{F}] = 0$ in $H_1(\mathcal{R})$—then the 2-chain that bounds $\mathcal{F}$ gives specific information about the cover. Intuitively, this 2-chain has the appearance of "filling in" $\mathcal{D}$ with triangles composed of projected 2-simplices from $\mathcal{R}$, as in Fig. 4.4. When translated into the language of algebraic topology, such a 2-chain is a relative 2-D homology class, a certain type of generator in $H_2(\mathcal{R}, \mathcal{F})$.

The following simple algebraic lemmas complete the setup.

Lemma 4.2 *Any nonzero 1-cycle $\zeta \in Z_1(\mathcal{F})$ defines a nonzero element of $H_1(\partial\mathcal{D})$.*

Proof By the definition of homology, $H_1(F) = Z_1(F)/B_1(\mathcal{F})$. However, $B_1(\mathcal{F}) = \partial(C_2(\mathcal{F})) = 0$, since $C_2(\mathcal{F}) = 0$ in the simplicial category; hence $Z_1(\mathcal{F}) = H_1(\mathcal{F}) = H_1(\partial\mathcal{D})$. □

Lemma 4.3 *A cycle $\zeta \in Z_1(\mathcal{F})$ is nonzero if and only if it has a nonzero coefficient at every fence edge.*

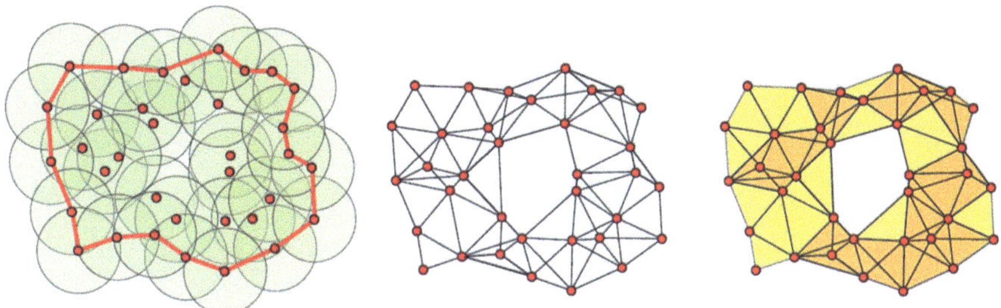

Fig. 4.3 In a sensor network with a sufficiently large hole in coverage (*left*), the communication graph (*center*) has a cycle that cannot be "filled in" by triangles. The filled in Rips complex (*right*) "sees" this hole, even as an abstract complex devoid of sensor node location data

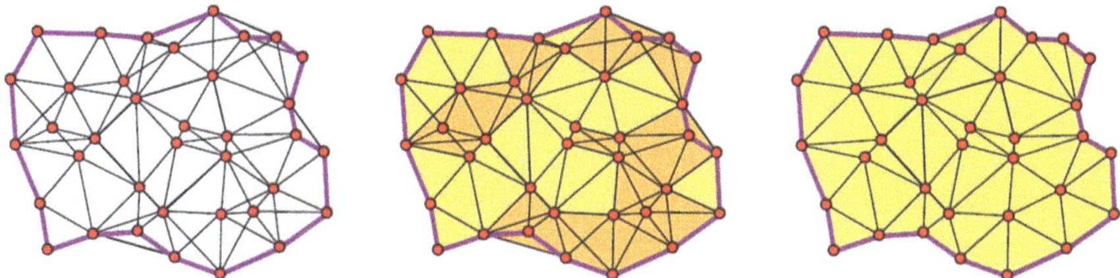

Fig. 4.4 The coverage criterion is an algebraic-topological formulation of the intuition of "filling in" the fence cycle $\mathcal{F}$ of the communication graph (*left*) with 2-simplices of the Rips complex $\mathcal{R}$ (*center*) so as to triangulate the domain $\mathcal{D}$ (*right*)

Proof If ζ is a cycle, then the coefficient of ζ at any pair of adjacent edges is the same up to a sign, because $\partial\zeta$ has coefficient zero at their common vertex. Since the boundary is connected, ζ has the same coefficient at every edge of $\mathcal{F}$ up to a sign. The lemma follows immediately. □

The following theorem is our principal coverage criterion.

Theorem 4.2 *For a set of nodes $\mathcal{X}$ in a domain $\mathcal{D} \subseteq \mathbb{R}^2$ satisfying Assumptions **A1–A4**, the sensor cover $\mathcal{U}_c$ contains $\mathcal{D}$ if there exists $[\alpha] \in H_2(\mathcal{R}, \mathcal{F})$ such that $\partial\alpha \neq 0$.*

For readers who struggle with the homological formalism, the example to keep in mind is that of a generator $[\alpha] \in H_2(\mathcal{R}, \mathcal{F})$ where α triangulates the domain $\mathcal{D}$ as in Fig. 4.4 (right).

We note (by Lemma 4.3) that the condition $\partial\alpha \neq 0$ can easily be evaluated by picking a single fence edge and testing whether the coefficient of $\partial\alpha$ on that edge is nonzero.

Proof We consider the simplicial realization map $\sigma: \mathcal{R} \mapsto \mathbb{R}^2$ which sends vertices of the abstract complex $\mathcal{R}$ to the corresponding node points of $\mathcal{X} \subseteq \mathcal{D}$ and which sends a k-simplex of $\mathcal{R}$ to the (potentially singular) k-simplex given by the convex hull of the vertices implicated. Via Assumption **A4,** σ takes the pair $(\mathcal{R}, \mathcal{F})$ to $(\mathbb{R}^2, \partial\mathcal{D})$; we therefore construct the following diagram from the long exact sequences:

$$\begin{array}{ccc} H_2(\mathcal{R}, \mathcal{F}) & \stackrel{\delta_*}{\rightarrow} & H_1(\mathcal{F}) \\ \downarrow \sigma_* & & \downarrow \sigma_* \\ H_2(\mathbb{R}^2, \partial\mathcal{D}) & \stackrel{\delta_*}{\rightarrow} & H_1(\partial\mathcal{D}) \end{array} \tag{4.1}$$

Here, δ_* acts on a class $[\alpha] \in H_2(\mathcal{R}, \mathcal{F})$ by taking the boundary: $\delta_*[\alpha] = [\partial\alpha] \in H_1(\mathcal{F})$. It follows from the naturality of the long exact sequence that the diagram of Relation 4.1 is commutative: $\delta_*\sigma_* = \sigma_*\delta_*$. The homology class $\sigma_*\delta_*[\alpha]$ is the winding number of $\partial\alpha$ about $\partial\mathcal{D}$.

By assumption, $\partial\alpha \neq 0$; hence, by way of Lemma 4.2, we observe that $\sigma_*\delta_*[\alpha] = \sigma_*[\partial\alpha] \neq 0$. By commutativity of Eq. 4.1, $\delta_*\sigma_*[\alpha] \neq 0$, and thus $\delta_*[\alpha] \neq 0$.

Assume that $\mathcal{U}$ does not contain $\mathcal{D}$ and choose $p \in \mathcal{D} - \mathcal{U}$. Since, by Lemma 4.1, every point in $\sigma(\mathcal{R})$ lies within $\mathcal{U}$, we have that $\sigma:(\mathcal{R}, \mathcal{F}) \mapsto (\mathbb{R}^2, \partial\mathcal{D})$ factors through the pair $(\mathbb{R}^2 - p, \partial\mathcal{D})$. However, $H_2(\mathbb{R}^2 - p, \partial\mathcal{D}) = 0$, as the following simple computation shows. Let $A = \mathbb{R}^2 - p$ and B be a small ball about p, so that $A \cap B$ is an open annulus homotopic to S^1. Let $A' = \partial\mathcal{D}$ and $B' = \emptyset$. Using the relative Mayer–Vietoris sequence, we have

$$\cdots \rightarrow H_2(S^1) \stackrel{\phi_*}{\rightarrow} H_2(\mathbb{R}^2 - p, \partial\mathcal{D}) \oplus 0 \stackrel{\psi_*}{\rightarrow} H_2(\mathbb{R}^2, \partial\mathcal{D}) \stackrel{\partial_*}{\rightarrow} H_1(S^1) \stackrel{\phi_*}{\rightarrow} \cdots \tag{4.2}$$

Since $(\mathbb{R}^2, \partial\mathcal{D})$ deformation retracts to the pair $\mathcal{D}, \partial\mathcal{D}$ fixing $\mathcal{D}$, we have that:

$$H_2(\mathbb{R}^2, \partial\mathcal{D}) \cong H_2(\mathcal{D}, \partial\mathcal{D}) \cong H_2(\mathcal{D}/\partial\mathcal{D}) \cong H_2(S^2) \cong \mathbb{R} \tag{4.3}$$

Since $p \in \mathcal{D}$, the homomorphism ∂_* takes the generator of $H_2(\mathbb{R}^2, \partial\mathcal{D})$ to that of $H_1(S^1)$. Equation 4.2 therefore simplifies to:

$$\cdots \rightarrow 0 \rightarrow H_2(\mathbb{R}^2 - p, \partial\mathcal{D}) \rightarrow \mathbb{R} \stackrel{\cong}{\rightarrow} \mathbb{R} \rightarrow \cdots \tag{4.4}$$

By exactness, $H_2(\mathbb{R}^2 - p, \partial\mathcal{D}) = 0$ and thus $\sigma_*[\alpha] = 0$: contradiction. □

Remark 4.2 This is not a sharp criterion. It is clearly possible to have the criterion always fail for injudicious choice of r_c. For example, if r_c is much larger than the bound in Assumption **A3**, then there will be many instances of coverage without a homological forcing. This being said, we note that even if one chooses the minimal acceptable bounds from Assumption **A3**, it is still possible to arrange the points to cover $\mathcal{D} - \mathcal{C}$ without the homological criterion detecting this, as illustrated in Fig. 4.5.

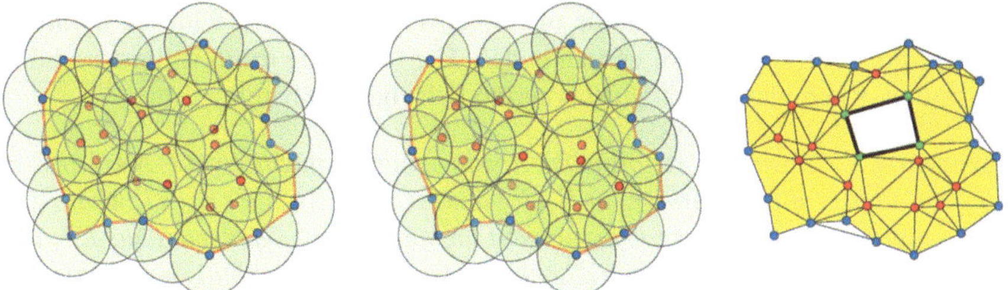

Fig. 4.5 Examples of two covers. The homological criterion holds for one (*left*) but not for the other (*center*), because of a 1-cycle in $\mathcal{R}$ (*right*). Note the fragility of the cover (*center*) within the 1-cycle: a small perturbation of the nodes creates a hole

4.4 Generators for Redundant Covers

Theorem 4.2 guarantees that the covering disks in fact cover the desired area. For reasons of power conservation, one would like to know which nodes could be "turned off" without impinging upon the coverage integrity. This is an important problem with a large literature, see, e.g., [22, 26]. A practical approach to this problem is implicit in homological methods.

Corollary 4.1 *If a homology class in $H_2(\mathcal{R}, \mathcal{F})$ satisfies the criterion of Theorem 4.2, then the restriction of $\mathcal{U}$ to those nodes which make up the representative α suffice to cover $\mathcal{D}$, for any choice of α in the homology class.*

Proof Let $\mathcal{U}^\alpha$ denote the restriction of $\mathcal{U}$ to the nodes implicated by the representative α. Assume that $\mathcal{U}^\alpha$ does not contain $\mathcal{D}$ and choose $p \in \mathcal{D} - \mathcal{U}^\alpha$. Lemma 4.1 implies that $\sigma(\mathcal{R}) \subseteq \mathcal{U}^\alpha$. Thus, $\sigma{:}(\mathcal{R}, \mathcal{F}) \mapsto (\mathbb{R}^2, \partial\mathcal{D})$ again factors through the pair $(\mathbb{R}^2 - p, \partial\mathcal{D})$, which has vanishing homology in dimension two. □

The independence of the choice of representative in the homology class is extremely important. If one chooses a "minimal" generator α—in the sense that α minimizes the number of 0-simplices within $[\alpha]$—then Corollary 4.1 yields a small subset of nodes which is guaranteed to cover the domain. Existing software packages for computing homology classes can "shrink" generators (though without rigor in terms of being truly minimal); hence, this is an implementable strategy. In Sect. 4.12, we give an example.

4.5 Hole Repair

Since the result of Theorem 4.2 is merely a criterion, one wishes to implement a strategy for guaranteeing coverage when the criterion fails. We present an elementary means for doing so via homology, the idea being to compute "minimal" generators in $H_1(\mathcal{R})$ so as detect holes. We consider a sensor network in which all nodes are initially in a "power saving" mode of low coverage radius r_c with the ability to increase the coverage radii of certain nodes. The following result is most useful in this setting, where the homological criterion fails, but just barely.

Theorem 4.3 *Consider a set of nodes $\mathcal{X}$ satisfying Assumptions **A1–A4**. Let $\Gamma = \{\gamma_i\}_1^K$ be a basis of K generators in $H_1(\mathcal{R})$ and let $N_i = ||\gamma_i||$ for each i, where $||\cdot||$ denotes length of the generator in terms of the number of nodes implicated. Let $\mathcal{U}'$ denote the set obtained from the collection $\mathcal{U}$ by enlarging all balls based at nodes in γ_i to balls of radius*

$$r_c'(i) = \frac{r_b}{2}\csc\frac{\pi}{N_i} \tag{4.5}$$

Then, $\mathcal{D} \subseteq \mathcal{U}.'$

Thus, for example, any Rips complex that has one or more "holes" of size four [as in Fig. 4.3 (right)], then the coverage region is guaranteed to contain $\mathcal{D}$ if we require $r_c \geq r_b\sqrt{2}$ for the implicated nodes defining where the hole is.

Proof The quantity $r_c'(i)$ represents the minimal radius needed to cover a regular N_i-gon. We claim that this is the limiting case.

Consider the image $\mathcal{L}_1 = \sigma(\gamma_i)$ of the loop γ_i in $\mathcal{D}$. This is a (not necessarily embedded) loop in $\mathcal{D}$. A point $x \in \mathcal{D}$ is enclosed by $\mathcal{L}_i$ if $[\mathcal{L}_i]$ is nonzero in $H_1(\mathbb{R}^2 - x) \cong \mathbb{Z}$ (this class is the *winding number* of the loop about x). We demonstrate that covering each node of γ_i with a ball of radius $r_c'(i)$ covers any such x. For such an x, it follows that one or more of the N_i edges of $\mathcal{L}$ subtends an angle at x of at least $2\pi/N_i$; for otherwise there would exist rays originating at x which miss $\sigma(\gamma_i)$ entirely, making $\mathcal{L}_i$ contractible in $\mathbb{R}^2 - x$ and the winding number zero. Let ab be such an edge. Taking cosines this inequality becomes:

$$\cos\frac{2\pi}{N_i} \geq \frac{|xa|^2 + |xb|^2 - |ab|^2}{2|xa||xb|} \geq 1 - \frac{r_b^2}{2|xa||xb|} \tag{4.6}$$

where we use the AM-GM inequality and the fact that $|ab| \leq r_b$ for the latter inequality. Since $\cos(2\pi/N_i) = 1 - 2\sin^2(\pi/N_i)$, we can rearrange to obtain $|xa||xb| \leq \left(r_c'(i)\right)^2$. Thus, x must lie within distance $r_c'(i)$ of the nearer of the two nodes a, b, as required.

We now create a modified complex $\mathcal{R}'$ obtained from $\mathcal{R}$ in the following manner. For each i, sew in an abstract 2-D disk along the loop γ_i. (If one wishes to remain in the simplicial category, one can triangulate the disk.) Next, extend the map σ to a continuous map $\sigma':\mathcal{R}' \mapsto \mathbb{R}^2$.

The long exact sequence yields a commutative diagram as in Eq. 4.1:

$$\begin{array}{ccccc} H_2(\mathcal{R}',\mathcal{F}) & \xrightarrow{\delta_*} & H_1(\mathcal{F}) & \xrightarrow{i_*} & H_1(\mathcal{R}') \\ \downarrow \sigma_*' & & \downarrow \sigma_*' & & \downarrow \sigma_*' \\ H_2(\mathbb{R}^2,\partial\mathcal{D}) & \xrightarrow{\delta_*} & H_1(\partial\mathcal{D}) & \xrightarrow{i_*} & H_1(\mathbb{R}^2) \end{array} \tag{4.7}$$

Because we have filled in all the generators of $H_1(\mathcal{R})$, we have that $H_1(\mathcal{R}') = 0$ and $\delta_*: H_2(\mathcal{R}', \mathcal{F}) \mapsto H_1(\mathcal{F})$ is onto. Exactness implies that there exists a generator $[\alpha]$ of $H_2(\mathcal{R}')$ with $\partial\alpha = \mathcal{F}$.

Assume by way of contradiction that there exists a point $p \in \mathcal{D} - \mathcal{U}'$. If $[\mathcal{L}_i] \neq 0 \in H_1(\mathbb{R}^2 - p)$ for any i, then $p \in \mathcal{U}'$ by the argument above. Therefore, assume that these homology classes vanish for all i. Since the set $\{\gamma_i\}$ forms a basis for $H_1(\mathcal{R})$, there exists a 2-chain ζ in $C_2(\mathcal{R})$ such that $\partial\zeta = \mathcal{F} - \sum_i c_i\gamma_i$ for some constants c_i. Applying σ to these 1-chains yields the equation $\partial\sigma(\zeta) = \partial\mathcal{D} - \sum_i c_i\mathcal{L}_i$. This descends to an equation in $H_1(\mathbb{R}^2 - p)$, since p is assumed to be not in $\mathcal{U}'$ and $\sigma(\zeta) \subseteq \mathcal{U} \subseteq \mathcal{U}'$ by Lemma 4.1. We know that $[\partial\mathcal{D}] \neq 0 \in H_1(\mathbb{R}^2 - p)$ since $p \in \mathcal{D}$. By assumption that all the winding numbers of $\mathcal{L}_i$ about p vanish, we have that $[\partial\sigma(\zeta)] \neq 0 \in H_1(\mathbb{R}^2 - p)$. However, $\zeta \in C_2(\mathcal{R})$ and is an algebraic sum of 2-simplices in $\mathcal{R}$. At least one such 2-simplex ς of ζ must therefore satisfy $sigma(\partial\varsigma) \neq 0 \in H_1(\mathbb{R}^2 - p)$, implying that $p \in$ sigma$(\zeta) \subseteq \mathcal{U} \subseteq \mathcal{U}'$. Contradiction.

It follows from this argument that, if one has the hardware constraint of a fixed coverage radius r_c which is larger that the bound $r_b/\sqrt{3}$, then one can get a better coverage criterion, as follows. Let N be the largest integer for which $r_c \leq 2r_b/\csc(\pi/N)$. Then, build a version of the Rips complex for the network which has all loops in the network of length less than or equal to N filled in by abstract 2-cells. Coverage is guaranteed if the resulting cell complex has a relative cycle in H_2 with non-vanishing boundary. □

4.6 Networks Without Boundaries

Among the conditions on the sensor networks to which these results apply, Assumptions **A1–A4** on the boundary are the least "natural" for a realistic network. In many contexts (real and hypothetical), networks are of large enough extent that boundary phenomena are ignorable. The homological criterion of Theorem 4.2 adapts to networks without boundaries in a number of possible ways: we outline the simplest such extension here.

Consider a cycle γ in the communication graph. One approach is to interrogate the network coverage with respect to this cycle: is the area bounded by this cycles projection to the plane covered? One must be careful: if the projection of γ to the plane is a simple closed curve, then it has a well-defined interior whose coverage can be queried via a homology computation. Cycles γ which have lots of self-intersection in the projection to the plane are generally to be avoided in a coverage querying context. Determining whether a given cycle in the network has a simple closed image is not trivial. The following simple (and well-known) criterion is efficacious.

Lemma 4.4 *Let γ be a 1-cycle in $\mathcal{R}$ whose span, $\langle\gamma\rangle$—the largest sub-complex of $\mathcal{R}$ generated by the nodes of γ—is precisely gamma. Then, the projection $\sigma(\gamma)$ of γ to the plane is a simple closed curve.*

Proof Assume that the images of two edges e_1 and e_2 of γ intersect in their interiors, forming an "X" in the plane. Since the lengths of these edges are no larger than r_b, it follows that at least one segment of this "X" from e_1 and one from e_2 have length no more than $\frac{1}{2}r_b$. The triangle inequality implies that two end vertices of these segments are within r_b, forming a new edge of $\langle\gamma\rangle$. □

Corollary 4.2 *For a planar network satisfying Assumptions **A1** and **A2**, choose a cycle γ with $\langle\gamma\rangle = \gamma$. If $H_2(\mathcal{R}, \gamma)$ has a generator $[\alpha]$ with $\partial\alpha \neq 0$, then the entire domain bounded by $\sigma(\gamma) \in \mathbb{R}^2$ lies within the cover $\mathcal{U}^\alpha$.*

Proof The argument of Theorem 4.2 suffices, thanks to Lemma 4.4. □

4.7 Domains with Arbitrary Planar Topology

Assumption A3 restricts the topology of the domain $\mathcal{D}$ in two features: connectivity of $\mathcal{D}$ and connectivity of $\partial\mathcal{D}$. It is not difficult to eliminate both of these requirements. If $\mathcal{D}$ is disconnected, then each connected component of $\mathcal{D}$ can be treated separately. If $\partial\mathcal{D}$ is disconnected, we can succeed if we have some extra information about the connected components of $\partial\mathcal{D}$.

Theorem 4.4 *Consider a set of nodes $\mathcal{X}$ satisfying Assumptions **A1–A4**, with **A3** modified as follows:*

Assumption A3′ *Nodes $\mathcal{X}$ lie in a compact connected domain $\mathcal{D} \subseteq \mathbb{R}^2$ whose boundary $\partial\mathcal{D}$ is piecewise-linear with vertices marked fence nodes $\mathcal{X}_f$. There is a partition of $\mathcal{X}_f$ into $\mathcal{X}_f^+ \sqcup \mathcal{X}_f^-$, representing those on the outer and inner boundary components, respectively.*

The sensor cover $\mathcal{U}_c$ contains $\mathcal{D}$ if there exists $[\alpha] \in H_2(\mathcal{R}, \mathcal{F})$ such that $\partial\alpha$ is nonzero on the outermost boundary component.

To evaluate the condition on α, we can pick any edge on the outermost boundary component and check whether $\partial\alpha$ has a nonzero coefficient at that edge (compare Lemma 4.3).

Proof This is a modification of the proof of Theorem 4.2. To start with, we can write the fence sub-complex as a disjoint union $\mathcal{F} = \mathcal{F}^+ \sqcup \mathcal{F}^-$, where $\mathcal{F}^+$ is the outermost fence component, and $\mathcal{F}^-$ is the union of the inner fence components. Similarly one can write $\partial\mathcal{D} = \partial^+\mathcal{D} \sqcup \partial^-\mathcal{D}$ for the domain boundary. The condition on α is then equivalent to the assertion that $\delta_*[\alpha] \neq 0$, where $\delta_*: H_2(\mathcal{R}, \mathcal{F}) \mapsto H_1(\mathcal{F}, \mathcal{F}^-) \cong H_1(\mathcal{F}^+)$ is the boundary map in the long exact sequence for the triple $(\mathcal{R}, \mathcal{F}, \mathcal{F}^-)$.

This time, we have a simplicial realization map $\sigma: (\mathcal{R}, \mathcal{F}, \mathcal{F}^-) \mapsto (\mathbb{R}^2, \partial\mathcal{D}, \partial - \mathcal{D})$, which gives us the following commutative diagram:

$$\begin{array}{ccccc} H_2(\mathcal{R}, \mathcal{F}) & \xrightarrow{\delta_*} & H_1(\mathcal{F}, \mathcal{F}^-) = & H_1(\mathcal{F}^+) \\ \downarrow \sigma_* & & \downarrow \sigma_* & \downarrow \sigma_* \\ H_2(\mathbb{R}^2, \partial\mathcal{D}) & \xrightarrow{\delta_*} & H_1(\partial\mathcal{D}, \partial^-\mathcal{D}) = & H_1(\partial^+\mathcal{D}) \end{array} \tag{4.8}$$

The equalities on the right of the diagram come from the excision theorem. Since $\sigma_*: H_1(\mathcal{F}^+) \mapsto H_1 \partial^+ \mathcal{D}$ is an isomorphism, the same is true of $\sigma_*: H_1(\mathcal{F}, \mathcal{F}^-) \mapsto H_1(\partial\mathcal{D}, \partial^-\mathcal{D})$.

Suppose there exists $[\alpha]$ satisfying the criterion in the theorem, so $\delta_*[\alpha] \neq 0$. By commutativity of Eq. 4.1 and since the middle map σ_* is an isomorphism, it follows that $\delta_*\sigma_*[\alpha] = \sigma_*\delta_*[\alpha] \neq 0$. Now assume, for a contradiction, that there is some point $p \in \mathcal{D} - \mathcal{U}$. Since it lies in $\mathcal{D}$, the point p is encircled by the outermost boundary component $\partial^+\mathcal{D}$ but not by any of the other boundary components. Since $p \notin \mathcal{U}$, the composite $\delta_*\sigma_*$ factors as:

$$H_2(\mathcal{R}, \mathcal{F}) \xrightarrow{\sigma_*} H_2(\mathbb{R}^2 - p, \partial\mathcal{D}) \xrightarrow{i_*} H_2(\mathbb{R}^2, \partial\mathcal{D}) \xrightarrow{\delta_*} H_1(\partial\mathcal{D}, \partial^-\mathcal{D}) \tag{4.9}$$

We claim that $\delta_* i_*: H_2(\mathbb{R}^2 - p, \partial\mathcal{D}) \mapsto H_1(\partial\mathcal{D}, \partial^-\mathcal{D})$ is the zero map, which gives the required contradiction since it implies that $\delta_*\sigma_*[\alpha] = 0$.

In fact, $\delta'_* = \delta_* i_*$ is the boundary map in the long exact sequence for the triple $(\mathbb{R}^2 - p, \partial\mathcal{D}, \partial^-\mathcal{D})$. Consider the following excerpt from that sequence:

$$\cdots \to H_2(\mathbb{R}^2 - p, \partial\mathcal{D}) \xrightarrow{\delta'_*} H_1(\partial\mathcal{D}, \partial^-\mathcal{D}) \xrightarrow{j_*} H_1(\mathbb{R}^2 - p, \partial^-\mathcal{D}) \to \cdots \tag{4.10}$$

By exactness, we can prove that $\delta'_* = 0$ by establishing instead that j_* is one-to-one. This can be read off from the following commutative diagram with exact rows, coming from the inclusion map of pairs $j: (\partial\mathcal{D}, \partial^-\mathcal{D}) \mapsto (\mathbb{R}^2 - p, \partial^-\mathcal{D})$.

$$\begin{array}{ccccc} \cdots & & H_1(\partial\mathcal{D}) & \xrightarrow{i_*} & H_1(\partial\mathcal{D}, \partial^-\mathcal{D}) \quad \cdots \\ \nearrow & & \downarrow i_* & & \downarrow j_* \\ H_1(\partial^-\mathcal{D}) & \xrightarrow{0} & H_1(\mathbb{R}^2 - p) & \xrightarrow{k_*} & H_1(\mathbb{R}^2 - p, \partial^-\mathcal{D}) \quad \cdots \end{array} \tag{4.11}$$

The geometric content here is that the map $H_1(\partial^-\mathcal{D}) \mapsto H_1(\mathbb{R}^2 - p)$ is zero, since the interior boundary cycles do not enclose p, whereas the map $H_1(\partial\mathcal{D}) \mapsto H_1(\mathbb{R}^2 - p)$ is onto since the outer boundary cycle does encircle p. It follows that the two maps labeled i_* have the same kernel and are both onto. By exactness, the map labeled k_* is one-to-one and therefore the same is true of j_*. This is what was required. □

It is not enough to have $\partial\alpha \neq 0$ as before. Consider the situation of Fig. 4.6, in which a small interior boundary component is a loop of four edges. Then, one can generate a relative 2-cycle consisting of the four boundary nodes along with a single interior node which is properly situated. This, of course, does not cover the domain.

We leave it to the reader to modify the statements of theorems in the following sections to accommodate the case of domains which for which connectivity or simple connectivity fail.

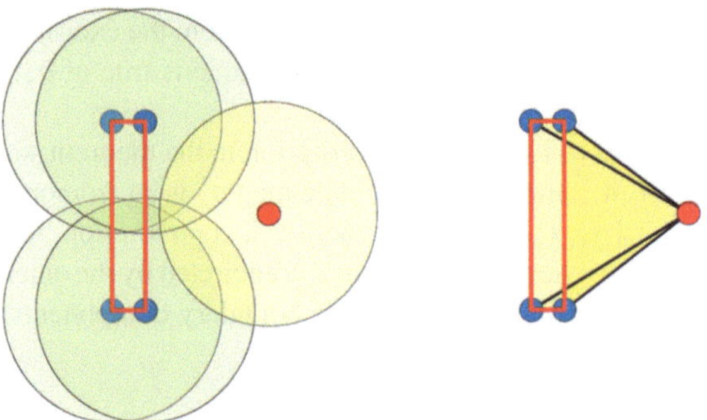

Fig. 4.6 An example of a small internal boundary component (*left*) giving rise to a fake relative 2-cycle (*right*) in the Rips complex

4.8 Opaque Boundaries and Communication Errors

We have not carefully specified the mechanism by which nodes communicate presence over a distance. From Assumption **A1**, it follows that communication signals are picked up purely as a function of distance, permeating the boundary of the domain if necessary. In certain physical situations, these communication signals may not be capable of boundary penetration (e.g., if they are visually detected beacons). One might wish to modify the assumptions with the following *opaque* boundary condition: *Each node can detect the identity of any node connected by a straight line in $\mathcal{D}$* of length at most r_b. One changes the Rips complex to include only those edges which communicate through unobstructed signals.

This is a particular example of the more general phenomenon of having communication errors of the form where two nodes within communication distance fail to establish a link. For the most general case, consider a system satisfying Assumptions **A1–A4** with Rips complex $\mathcal{R}$. Define a Rips complex with omissions, $\mathcal{ER}$, to be any sub-complex of $\mathcal{R}$ containing $\mathcal{F}$ (we assume perfect control of the fence nodes). This $\mathcal{ER}$ may result as a random error in establishing communication links or, as above, as a systematic failure to establish links near certain types of boundaries.

Theorem 4.5 *Consider a set of nodes $\mathcal{X}$ in a domain $\mathcal{D} \subseteq \mathbb{R}^2$ satisfying assumptions **A1–A4** with $\mathcal{ER}$ a Rips complex with omissions. The sensor cover $\mathcal{U}_c$ contains $\mathcal{D}$ if there exists $[\alpha] \in H_2(\mathcal{ER}, \mathcal{F})$ such that $\partial\alpha \neq 0$.*

Proof Since $\mathcal{ER} \subseteq \mathcal{R}$, we have:

$$\begin{array}{ccc} H_2(\mathcal{ER}, \mathcal{F}) & \stackrel{\delta_*}{\longrightarrow} & H_1(\mathcal{F}) \\ \downarrow \sigma_* & & \downarrow \sigma_* \\ H_2(\mathbb{R}^2, \partial\mathcal{D}) & \stackrel{\delta_*}{\longrightarrow} & H_1(\partial\mathcal{D}) \end{array} \tag{4.12}$$

The remainder of the proof follows exactly as in Theorem 4.2. □

This result implies that the homological coverage criterion relies on the coarse metric data of Assumption **A1** only in the positive sense. The criterion does not use the fact that a failure to communicate implies a lower bound on the distance between nodes.

4.9 Variable Radii

Assumptions **A1** and **A2** on the radial symmetry of sensors are physically unrealistic: a more accurate model would incorporate asymmetry and/or variable radii, to accommodate errors or fluctuations in signals. It is possible to apply the homological criterion to systems with asymmetric broadcast domains using the Rips complex with omissions of Sect. 4.8. One chooses r_b to be an upper bound for the broadcast signal distance and $r_c \geq r_b/\sqrt{3}$. The communication network then establishes links between certain nodes, but not purely as a function of distance. While this method is applicable, there is a wastefulness in the bound on rc in terms of the maximal broadcast distance.

We therefore consider systems whose radii r_c and r_b vary from node to node, as a next step toward dealing with asymmetry in sensor networks. Consider the case where a system of nodes $\mathcal{X} = \{x^i\}$ satisfies a modified set of assumptions:

Assumption V1 Nodes $\mathcal{X} = \{x^i\}$ broadcast their unique ID numbers. The identity of each node can be detected by any node within its broadcast radius r_b^i.

Assumption V2 Nodes have radially symmetric covering domains of cover radius $r_c^i \geq r_b^i/\sqrt{3}$.

Assumption V3 Nodes $\mathcal{X}$ lie in a compact connected domain $\mathcal{D} \subseteq \mathbb{R}^2$ whose boundary $\partial\mathcal{D}$ is connected and piecewise-linear with vertices marked fence nodes $\mathcal{X}_f$.

Assumption V4 Each fence node $v \in \mathcal{X}_f$ knows the identities of its neighbors on $\partial\mathcal{D}$ and these neighbors both lie within distance r_b^i of v.

We modify the construction of the Rips complex as follows. For any pair of nodes x^i and x^j, there is an edge in $\mathcal{R}$ if and only if the distance between x^i and x^j in $\mathcal{D}$ is less than or equal to the minimum of r_b^i and r_b^j. The full complex $\mathcal{R}$ is then the maximal simplicial complex for the edge set as defined. The fence cycle $\mathcal{F}$ is defined in the same way as before, with vertex set $\mathcal{X}_f$ and an edge between each pair of adjacent nodes along the fence. We define the variable-radius cover $\mathcal{U}_c$ in this context to be the union of closed disks of radii r_c^i centered at node x^i.

Theorem 4.6 *For a set of nodes $\mathcal{X}$ in a domain $\mathcal{D} \subseteq \mathbb{R}^2$ satisfying the variable-radius Assumptions* ***V1–V4****, the variable-radius cover $\mathcal{U}_c$ contains $\mathcal{D}$ if there exists $[\alpha] \in H_2(\mathcal{R}, \mathcal{F})$ such that $\partial\alpha \neq 0$.*

Proof The proof of Theorem 4.2 being topological is largely independent of the geometry of the system. The crucial geometric step is in the application of Lemma 4.1. We now verify that the variable-radius version of this lemma holds.

Consider a triple of points $\{x_1, x_2, x_3\}$ which span a triangle in $\mathcal{R}$ with side lengths l_{12}, l_{13}, and l_{23}, where $l_{ij} \leq \min(r_d^i, r_d^j)$. We must show that the three disks of radius r_b^i centered on x_i meet at a common point (and hence cover the triangle spanned by x_1, x_2, x_3).

Consider the continuous function:

$$f(x) = \max_{i=1}^{3} f_i(x) = \max_{i=1}^{3} \frac{||x - x_i||}{r_d^i}$$

Since $f(x) \to \infty$ as $||x|| \to \infty$ the function attains a global minimum, say $\lambda = f(x_0)$. We must show that $\lambda \leq 1/\sqrt{3}$.

The minimizer x_0 must lie inside the triangle $x_1x_2x_3$, because any point x outside the triangle can be perturbed so as to decrease all three distances $||x - x_i||$ simultaneously. In more detail, this argument shows that x_0 lies within the convex hull of its *critical vertices*, defined as those vertices x_i for which $f(x_0) = f_i(x_0)$.

There are two cases. If x_0 has two critical vertices x_i, x_j, then x_0 lies on the edge x_ix_j and $\lambda = f_i(x_0) = f_j(x_0) = l_{ij}/(r_d^i + r_d^j) \leq 1/2$, which is less than $1/\sqrt{3}$. Otherwise, all three vertices x_1, x_2, x_3 are critical. The largest of the three angles $\theta_{ij} = \angle x_ix_0x_j$ satisfies $\theta_{ij} \geq 2\pi/3$. The interior bisector of this angle meets the edge x_ix_j at a point y which divides the edge in the ratio $||x0 - x_i||$:$||x0 - x_j||$ or r_i:r_j. Using the sine rule for triangle x_0yx_i we then have:

$$\lambda r_i = ||x_0 - x_i|| = ||y - x_i|| \cdot \frac{\sin \angle x_0, yx_i}{\sin \frac{\theta_{ij}}{2}} \leq \frac{l_{ij}r_i}{r_i + r_j} \cdot \frac{1}{\sin \frac{\pi}{3}} \leq \frac{r_3}{\sqrt{3}}$$

giving the required bound.

The proof of the theorem now follows that of Theorem 4.2 precisely. □

Of course, the results on minimal generators and Rips complexes with omissions still apply in this setting as well, as the reader may check.

4.10 Barrier Coverage in 3D

We consider the following modification of the physical workspace of the nodes. Let the nodes be points in a 3-D tube of the form $\mathcal{D} \times \mathbb{R}$ for $\mathcal{D} \subseteq \mathbb{R}^2$ as in Assumption **A3**, and let the fence nodes lie in $\mathcal{D} \times \{0\}$ and satisfy **A4**. We define $\mathcal{U} \subseteq \mathbb{R}^2 \times \mathbb{R}$ by placing a 3-D ball of radius r_c at each $x_i \in \mathcal{X}$. The problem of *barrier coverage* is to determine whether there is a path connecting $\mathcal{D} \times \{-\infty\}$ to $\mathcal{D} \times \{+\infty\}$ avoiding $\mathcal{U}$: see Fig. 4.7.

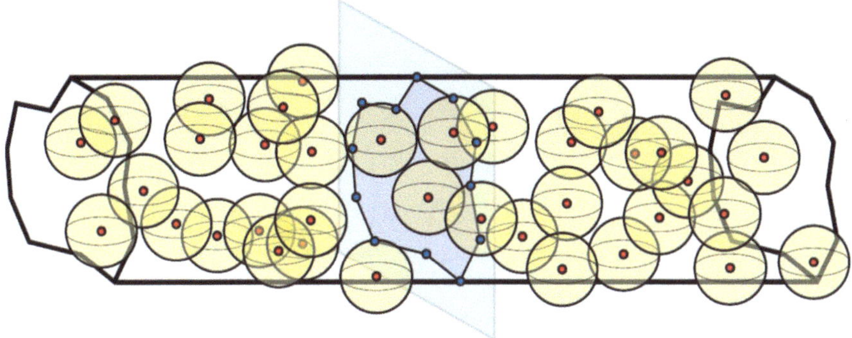

Fig. 4.7 Barrier coverage in a 3-D tube means the non-existence of a path from one end of the tube to the other avoiding 3-D balls of coverage about the nodes. The vestige of the fence cycle $\mathcal{F}$ is a cycle of nodes about the meridian $\partial\mathcal{D} \times \{0\}$ (balls of coverage not drawn along $\mathcal{F}$ for reasons of clarity)

We construct a Rips complex as before, connecting nodes if they are within distance r_b in $\mathcal{D} \times \mathbb{R}$. From **A4**, it follows that the fence cycle $\mathcal{F}$ is precisely $\partial\mathcal{D} \times \{0\}$. Our homological criterion immediately yields a criterion for barrier coverage.

Theorem 4.7 *A collection of nodes in $\mathcal{D} \times \mathbb{R}$ satisfying **A1–A4** as above has barrier coverage if there exists $[\alpha] \in H_2(\mathcal{R}, \mathcal{F})$ with $\partial\alpha \neq 0$.*

Proof We prove a stronger result in the spirit of Corollary 4.1. The proof of Lemma 4.2 holds for the 2-skeleton of the Rips complex: Three points determine a plane that intersects the balls in disks of radius r_c. Hence, the simplicial realization map $\sigma{:}\mathcal{R} \mapsto \mathcal{D} \times \mathbb{R}$ takes any 2-cycle α to a subset of $\mathcal{U}^\alpha$, the cover restricted to the nodes of α.

Let $\pi{:}\mathbb{R}^2 \times \mathbb{R} \mapsto \mathbb{R}$ denote projection to the second factor. Assume that $p{:}\mathbb{R} \mapsto \mathcal{D} \times \mathbb{R} - \mathcal{U}^\alpha$ is a continuous curve with $\lim_{x \to \pm\infty} \pi \circ p(x) = \pm\infty$. Since every point in $\sigma(\alpha)$ lies within $\mathcal{U}^\alpha$, we have that $\sigma{:}(\alpha, \partial\alpha) \mapsto (\mathbb{R}^2 \times \mathbb{R}, \partial\mathcal{D} \times \{0\})$ factors through the pair $(\mathbb{R}^2 \times \mathbb{R} - p, \partial\mathcal{D} \times \{0\})$. However, let $A = (\mathbb{R}^2 \times \mathbb{R}) - p$ and B be a neighborhood of p, so that $A \cap B$ is an annular tube homotopic to S^1. Let $A' = \partial\mathcal{D} \times \{0\}$ and $B' = \emptyset$. Consequently, we have

$$\begin{aligned} &\to H_2(S^1) \xrightarrow{\phi_*} H_2((\mathbb{R}^2 \times \mathbb{R}) - p, \partial\mathcal{D} \times \{0\}) \\ &\oplus\ 0 \xrightarrow{\psi_*} H_2((\mathbb{R}^2 \times \mathbb{R}), \partial\mathcal{D} \times \{0\}) \xrightarrow{\partial_*} H_1(S^1) \to \end{aligned} \tag{4.13}$$

Since $H_2((\mathbb{R}^2 \times \mathbb{R}), \partial\mathcal{D} \times \{0\}) \cong H_2(\mathcal{D}, \partial\mathcal{D}) \cong \mathbb{R}$ and ∂_* is an iso-morphism, we obtain:

$$\cdots \to 0 \to H_2((\mathcal{D} \times \mathbb{R}) - p, \partial\mathcal{D} \times \{0\}) \to \mathbb{R} \xrightarrow{\cong} \mathbb{R} \to \cdots \tag{4.14}$$

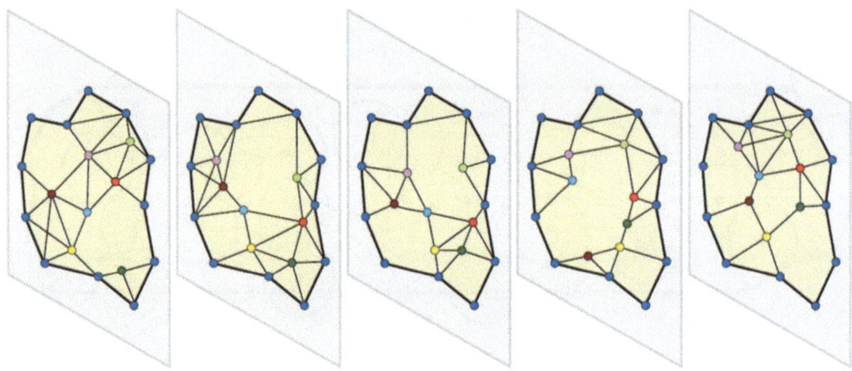

Fig. 4.8 A mobile network with fixed fence nodes sampled at five time segments: can an evader avoid being caught in the time-dependent union of coverage disks?

By exactness, $H_2((\mathbb{R}^2 \times \mathbb{R}) - p, \partial\mathcal{D} \times \{0\}) = 0$ and thus, $\sigma_*[\alpha] = 0$: contradiction. □

4.11 Pursuit-Evasion and Mobile Nodes

Consider a situation in which the node positions are a continuous function of time: $\mathcal{X} = \mathcal{X}_t \subseteq \mathcal{D}$ for $t \in [0, 1]$. Assume that the network is sampled to give a finite sequence of connectivity graphs $\{\Gamma_i\}_0^N$ at times $0 = t_0 < \cdots < t_N = 1$, as in Fig. 4.8. We assume the following:

Assumption T:1 If two nodes are connected at time steps t_i and t_{i+1}, then they remain within the broadcast radius r_b for all $t_i \leq t \leq t_{i+1}$.

Assumption T:2 Nodes may go off-line or come on-line, represented by deleting the nodes in the appropriate graph Γ_i.

Assumption T:3 Fence nodes always remain fixed and on-line.

We now address the question of whether there can be a "wandering" loss of coverage. It may be the case that at no time $t \in [0, 1]$ does there exist a complete sensor coverage of the domain; however, the changes may obstruct any sequence of points from "jumping" from one hole to the next, avoiding the coverage domain. Verifying the lack of wandering holes is a particular type of pursuit-evasion problem with relevance to problems in security and defense. Note that this problem is distinct from the "sweeping" coverage problem, in which one wants to know whether the union of the cover sets $\bigcup_t \mathcal{U}(t)$ contains $\mathcal{D}$.

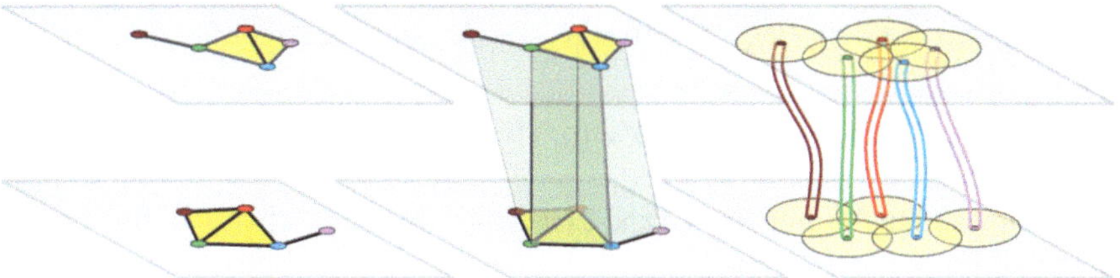

Fig. 4.9 Subsequent Rips complexes (*left*) are attached via prisms between matching simplices (*center*) to capture the topology of the mobile cover (*right*)

4.11.1 A Prism Complex

We present a homological criterion for guaranteeing no wandering holes via computing the homology of a certain space derived from the sequence of Rips complexes $\mathcal{R}_i$.

Definition 4.2 Given a sequence $\{\Gamma_i\}$ of vertex-labeled communication graphs as above, define the stacked Rips complex $\mathcal{SR}$ to be the cell complex obtained from the disjoint union $\bigsqcup_i \mathcal{R}_i$ of the Rips complexes $\mathcal{R}_i$ of Γ_i by the following operation:

For each k-simplex $[v_{\alpha_1}, \dots, v_{\alpha_{k+1}}]$ of $\mathcal{R}_i$ which is also a k-simplex on the same vertices in $\mathcal{R}_{i+1}$, connect these k-simplices by a prism $\Delta^k \times [0, 1]$ with $\Delta^k\{0\}$ glued to $\mathcal{R}_i$ and $\Delta^k \times \{1\}$ glued to $\mathcal{R}_{i+1}$.

We treat the time variable $t \in [0, 1]$ as an extra dimension and consider the problem of evasive coverage in $\mathcal{D} \times [0, 1]$. The complex $\mathcal{SR}$ has a natural "prism" structure: $\mathcal{SR}$ is a 1-parameter family of simplicial Rips complexes indexed by $t \in [0, 1]$, these "slices" being equal to $\mathcal{R}_i$ at t_i. See Fig. 4.9. We likewise consider the moving covers as a 1-parameter family in a 3-D setting. If $\mathcal{U}_t$ denotes the radius r_c cover of nodes $\mathcal{X}_t$ at time t, embed the time-varying covers into $\mathcal{D} \times [0, 1]$ via $\mathcal{U}_t \subseteq \mathcal{D} \times \{t\}$. The problem of wandering loss of coverage now becomes the question of whether the complement of the union $\bigcup_t \mathcal{U}_t$ in $\mathcal{D} \times [0, 1]$ has a "tunnel" running from bottom ($t = 0$) to top ($t = 1$).

Theorem 4.8 *Consider a time-varying set of nodes $\mathcal{X}_t$ in a domain $\mathcal{D} \subseteq \mathbb{R}^2$ satisfying Assumptions* ***A1–A4*** *and* ***T1–T3****. Then, for any continuous curve $p{:}[0, 1] \mapsto \mathcal{D}$, $p(t)$ must lie in $\mathcal{U}_t$ for some $0 \leq t \leq 1$ if there exists $[\alpha] \in H_2(\mathcal{SR}, \mathcal{F} \times [0, 1])$ such that $\pi_*(\partial\alpha) \neq 0$, where $\pi{:}\mathcal{F} \times [0, 1] \mapsto \mathcal{F}$ is the projection map.*

Proof As in the proof of Theorem 4.2, we consider a simplicial realization map $\sigma{:}\mathcal{SR} \mapsto \mathbb{R}^2 \times [0, 1]$. Define σ as follows. Given the structure of $\mathcal{SR}$ as a family of Rips complexes $\mathcal{R}_t$ indexed by $t \in [0, 1]$, let σ send each slice to $\sigma(\mathcal{R}_t) \subseteq \mathcal{D} \times \{t\}$, where σ is the realization map from the proof of Theorem 4.2 and the vertices are sent to $\mathcal{X}_t$.

The map σ takes the pair $(\mathcal{SR}, \mathcal{F} \times [0, 1])$ to $(\mathbb{R}^2 \times [0, 1], \partial\mathcal{D} \times [0, 1])$, yielding the following diagram:

$$\begin{array}{ccc} H_2(\mathcal{SR}, \mathcal{F} \times [0,1]) & \xrightarrow{\delta_*} & H_1(\mathcal{F} \times [0,1]) \\ \downarrow \bar{\sigma}_* & & \downarrow \bar{\sigma}_* \\ H_2(\mathbb{R}^2 \times [0,1], \partial\mathcal{D} \times [0,1]) & \xrightarrow{\delta_*} & H_1(\partial\mathcal{D} \times [0,1]) \end{array} \tag{4.15}$$

It follows from Assumption **T3** and Lemma 4.2 that $\pi_*\sigma_*\delta_*[\alpha] \neq 0$. By commutativity of Eq. 4.1, $\sigma_*[\alpha] \neq 0$.

Assume that there exists a continuous curve $p{:}[0,1] \mapsto \mathcal{D} \times [0,1]$ of points $p(t) \in \{\mathcal{D} \times \{t\} - \mathcal{U}_t\}$. We claim that $\sigma(\mathcal{SR}) \subseteq \bigcup_t \mathcal{U}_t$. Assume that the nodes $\{x_i(t)\}_{i=1}^{k+1}$ span a k-simplex of $\mathcal{R}_t \subseteq \mathcal{SR}$ at some fixed time t. Then, σ sends this to the convex hull of these nodes in $\mathbb{R}^2 \times \{t\}$. From Definition 4.2 and Assumption **T1**, any edge in $\mathcal{R}_t$ implies that the node points implicated by this edge are within distance r_b at time t. An application of Lemma 4.1 then guarantees that the convex hull of these nodes lies within $\mathcal{U}_t$.

We conclude from this and the existence of the wandering curve p that $\sigma{:}(\mathcal{SR}, \mathcal{F} \times [0,1]) \mapsto (\mathbb{R}^2 \times [0,1], \partial\mathcal{D} \times [0,1])$ factors through the pair $(\mathbb{R}^2 \times [0,1] - p, \partial\mathcal{D} \times [0,1])$. However, this has vanishing H_2, using the same argument as in Theorem 4.7. Thus, $\sigma_*[\alpha] = 0$: contradiction. □

4.11.2 A Simplicial Model

In practice, computing with the stacked Rips complex is inconvenient. The software we use is meant for simplicial complexes, not the more general prism complex $\mathcal{SR}$. We therefore provide a simple means of reducing the stacked Rips complex to a simplicial object which is much smaller and simpler to encode.

Definition 4.3 Given a collection of network graphs $\{\Gamma_i\}$ as in Definition 4.2, define the amalgamated Rips complex to be the space obtained from the disjoint union $\bigsqcup_i \mathcal{R}_i$ of the Rips complexes $\mathcal{R}_i$ of Γ_i by the following operation:

For each k-simplex $[v_{\alpha_1}, \ldots, v_{\alpha_{k+1}}]$ of $\mathcal{R}_i$ which is also a k-simplex on the same vertices in $\mathcal{R}_{i+1}$, identify these simplices.

A few observations are in order. First, the amalgamated Rips complex $\mathcal{AR}$ is a cell complex built from simplices. It is not, properly speaking, a (combinatorial) simplicial complex since there may be, e.g., more than one 1-simplex connecting two vertices; hence, cells in this complex are not uniquely defined by their faces. Second, since the fence nodes are assumed stationary, the fence cycle $\mathcal{F}$ is fixed in each $\mathcal{R}_i$ and thus is identified to yield a well-defined cycle $\mathcal{F} \subseteq \mathcal{AR}$.

Proposition 4.1 *The pair* $(\mathcal{SR}, \mathcal{F} \times [0,1])$ *is homotopy equivalent to* $(\mathcal{AR}, \mathcal{F})$.

Proof For each i, consider the maximal sub-complex $\mathcal{S}_i \subseteq \mathcal{R}_i$ which is also a sub-complex of $\mathcal{R}_{i+1}$. The prism sub-complex $\mathcal{S}_i \times [0,1] \subseteq \mathcal{SR}$ is a properly embedded sub-complex; hence the collapse of $\mathcal{S}_i \times [0,1]$ to the simplicial sub-complex $\mathcal{S}_i$ in $\mathcal{AR}$ is a homotopy equivalence. The amalgamated complex $\mathcal{AR}$ is the result of applying the sequence of collapses to $\mathcal{SR}$, and the sub-complex $\mathcal{F} \times [0, 1] \subseteq \mathcal{SR}$ is collapsed via projection of the second factor. □

This immediately implies the following:

Corollary 4.3 *The homological condition of Theorem 4.8 is satisfied if and only if $H_2(\mathcal{AR}, \mathcal{F})$ has a generator $[\alpha]$ with $\partial\alpha \neq 0$.*

These hypotheses are preferable to those of Theorem 4.8 in that the spaces involved are smaller, simplicial, and there is no condition involving the projection of the boundary of the generator. For a software package that can handle only true combinatorial simplicial complexes, there is a simple modification of $\mathcal{AR}$ available. Since the homological criterion resides in H_2, one can identify all k-simplices with the same boundary for $k \geq 2$. Only the multiple 1-simplices need be distinguished, and these may be handled by inserting additional vertices and refining the cell structure.

4.12 Computation

Unlike homotopy groups (such as the fundamental group π_1), homology is computable, and existing software packages make the homological coverage criteria of this chapter implementable for reasonable numbers of nodes. We have used the open-source package Plex [33], which consists of: (1) C++ code for manipulating simplicial complexes, written by Patrick Perry; (2) C++ code for persistent homology calculations, written by Lutz Kettner and Afra Zomorodian, published independently as part of the CGAL project [7]; (3) a MATLAB front-end and script library, designed and written by Vin de Silva and Patrick Perry.

Since we use pre-existing code for homology computations, a few remarks are in order with regard to implementation.

1. Plex does not automatically compute relative homology. In order to compute homology relative to the fence, we use the following simple procedure. To compute $H_2(\mathcal{R}, \mathcal{F})$, add a disjoint abstract vertex to $\mathcal{R}$ and augment this vertex to every simplex in $\mathcal{F}$. This is called placing a cone over the sub-complex $\mathcal{F}$, and it yields a complex $\mathfrak{C}(\mathcal{R}, \mathcal{F})$ whose homotopy type is that of the quotient space $\mathcal{R}/\mathcal{F}$. It follows from the Excision Theorem [21] and homotopy invariance that $H_*(\mathcal{R}, \mathcal{F}) \cong H_*(\mathcal{R}/\mathcal{F}) \cong H_*(\mathfrak{C}(\mathcal{R}, \mathcal{F}))$ for $* \geq 1$; hence, this faithfully captures the homology.
2. Common exposition of homology phrases everything in terms of linear algebra on real vector spaces, for clarity and intuition. In general, homology can be computed with any coefficient ring. The real coefficients that we use for intuition are not optimal for computation, since round-off error can impact computation. To avoid round-off error, we use homology with coefficients in the field $\mathbb{Z}_2$. All of our arguments are independent of the field coefficients used; hence the criterion is still valid with this assumption.

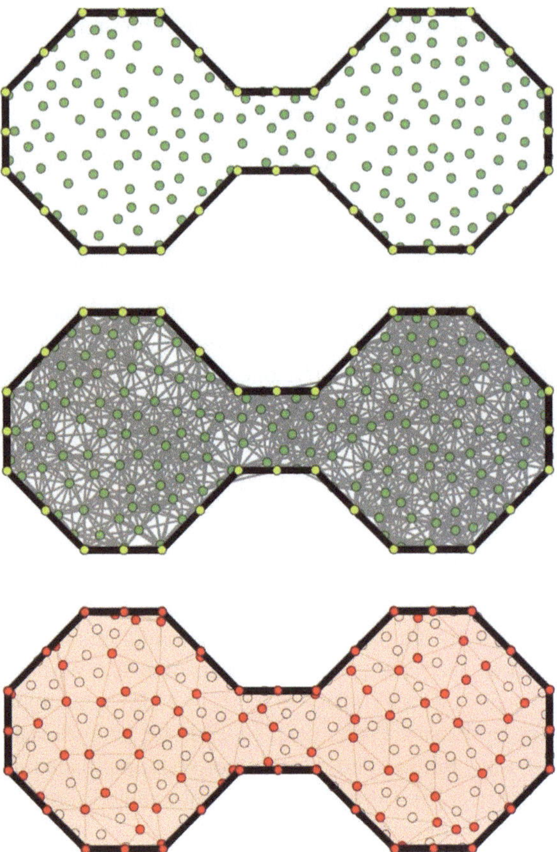

Fig. 4.10 A typical simulation: (*top*) the locations of 212 nodes in $\mathcal{D}$; (*center*) the image of the Rips complex $\mathcal{R}$ projected to $\mathcal{D}$; (*bottom*) a simple generator of $H_2(\mathcal{R}, \mathcal{F})$ extracts 101 nodes which are guaranteed to cover $\mathcal{D}$, leaving 111 nodes to be safely put into sleep mode

3. We compute generators for homology using the persistent homology algorithm, with the interior simplices being processed first and the cone simplices being processed last. Under this ordering, the algorithm is guaranteed to give a unique homology cycle spanning the fence if any exists (although this uniqueness does not seem to be significant). The cycle can be read off explicitly from the results of the computation.

Figure 4.10 shows a network in a simply connected domain with 212 nodes that satisfies the homological coverage criterion of Theorem 4.2. The figure also shows the image of the Rips complex in $\mathbb{R}^2$ under the realization map σ. A choice of a "simple" generator shows that 111 of the nodes may be put in sleep mode without a loss of coverage. Of necessity, this illustration shows the location of the nodes within the domain. We stress that the algorithms have no knowledge of this data. The input to the problem is the network connectivity graph and the fence cycle in that graph. The generator shown here is the one produced by the homology computation, with no subsequent optimization. No other geometric data are used.

We do not at this time present a complete analysis of the numerical implementation of the coverage criterion.

4.13 Summary and Outlook

The applicability of homology theory to sensor networks initiated in this chapter is not as surprising as might at first appear. Indeed, the two fields share several features. Problems in both homology and sensor networks have as inputs a large collection of local objects (simplices, sensors) with local interaction rules (faces, communication). From this collection (chain complex, sensor network), one seeks to determine global properties of the system (homology, coverage). The primary point of departure is that chain complexes carry with them a rich algebraic structure which can be exploited to great effect. We have demonstrated that certain features of this algebraic structure carry over to answer important questions in coverage, power conservation, and evasion-detection. This represents a new and powerful importation of algebraic tools in networks.

4.13.1 Remarks

1. We have not specified communication protocols on the level of hardware, having concerned ourselves in this chapter with the mathematical tools. We claim, however, that the Rips complex can be built in a distributed fashion on the hardware level: see [32]. We expect the signal complexity of this operation to be reasonable, since the Rips complex is completely determined by its 1-skeleton.
2. In this chapter, we have focused on the case where there is complete control over the fence nodes. In practice, such control may not be available. By endowing nodes with the capability of detecting the boundary of the domain, it is possible to reconstruct a fence sub-complex $\mathcal{F}$ composed of nodes near the boundary. Since these are not assumed to be well spaced (as in **A4**), the proofs of all the results here are invalid. We demonstrate in [11, 12] how to recover some of the results of this chapter in that more general case via *persistent homology*.
3. We stress that the coverage criterion is not if and only if. It is a rigorous test to guarantee coverage, and thus, any system which is "just barely" covered will likely fail that test.
4. The test as given in this chapter is centralized: a distributed coverage algorithm is greatly desirable.

4.13.2 Questions

This chapter represents merely the first step in applications of algebraic topology to sensor networks. We comment on possible and probable extensions below.

1. What is the computational complexity of the homological criterion as a function of number of nodes? The most straightforward algorithm for computing homology (using Smith normal form) can be quintic in the number of simplices. More recent algorithms are much faster, but the sub-quadratic algorithm of [13] relies on duality for Euclidean spaces and is not applicable for arbitrary simplicial complexes. Our experiments hint at a sub-quadratic run-time, and it may be that Rips complexes of planar networks have a sufficiently restricted topology to merit such a claim.
2. Can one construct an effective homological coverage criterion which is distributed, allowing nodes with limited computational capabilities to compute local homology? What are the demands on the nodes' computational power and memory in such a system? What demands are made on the communication network in a distributed homology computation?
3. Can the mobile-network coverage criterion for wandering holes be made asynchronous? Rather than sampling the entire network at once, subsets of nodes should sample their connectivity and register their network graph with a central processor. Does a homological criterion holds for such systems?
4. By changing the bound in **A2** to $r_c \geq r_b$, the homological criterion verifies 3-coverage in a planar network (a simple exercise). Is it possible to verify k-coverage for any k via homology? One wants to impose as few restrictions on rc as possible.
5. In practice, coverage and communication domains are not radially symmetric: elliptical or conical shapes are closer to reality in many cases. Is it possible to construct a homological coverage criterion for sensors whose communication and/or coverage domains are not radially symmetric? What additional capabilities do the sensors require in order to handle such asymmetry?
6. With the exception of the work in Sect. 4.11, we are working in a setting for which it is desired that there are more than enough sensors necessary to cover the domain. In such a sensor-rich environment, it is possible for the Rips complex to attain a very high dimension. This is highly undesirable for computational reasons. Is there a way to compress the Rips complex in a preprocessing step without changing the appropriate homology group? This seems reasonable: a 20-dimensional simplex implies a cluster of nodes, most of which should be redundant.
7. If we endow the nodes with additional capabilities, such as the ability to measure some angular data about neighboring nodes, what global problems can be solved? Problems involving degree computation and target isolation are solvable with only a very weak form of angular data at the nodes [19].
8. The sensor networks of this chapter are relatively idealized. Real sensors and real networks have unavoidable stochastic features. Is it possible to develop a

homology theory with "stochastic simplices" which returns rigorous coverage criteria in the form of, perhaps, "expected" homology classes?

References

1. M. Allili, K. Mischaikow, A. Tannenbaum, Cubical homology and the topological classification of 2D and 3D imagery, in IEEE International Conference on Image Processing (2001), pp. 173–176
2. A. Ames, A homology theory for hybrid systems: hybrid homology. Lect. Notes Comput. Sci. **3414**, 86–102 (2005)
3. M. Batalin, M. Hattig, G. Sukhatme, Mobile robot navigation using a sensor network, in Proceddings of ICRA (2004)
4. M. Batalin, G. Sukhatme, Spreading out: a local approach to multi-robot coverage, in Proceedings of 6th International Symposium on Distributed Autonomous Robotic Systems, Fukuoka, Japan (2002)
5. K. Bekris, L. Kavraki, A review of recent results in robotic sensor networks, in ACM Computing Surveys, vol. V(N) (2005)
6. N. Bulusu, J. Heidelmann, D. Estrin, Adaptive beacon placement, in Proceedings of the Conference on Distributed Computing Systems (2001)
7. Computational Geometry Algorithms Library, http://www.cgal/org/
8. P. Corke, R. Peterson, D. Rus, Localization and navigation assisted by cooperating networked sensors and robots. Int. J. Robot. Res. **24**(9), 771 ff (2005)
9. J. Cortes, S. Martinez, T. Karatas, F. Bullo, Coverage control for mobile sensing networks. IEEE Trans. Robot. Autom. **20**(2), 243–255 (2004)
10. V. de Silva, G. Carlsson, Topological estimation using witness complexes, in Symposium Point-Based Graphics, ETH Zurich (2004)
11. V. de Silva, R. Ghrist, Coverage in sensor networks via persistent homology. Alg. Geom. Topology (2007)
12. V. de Silva, R. Ghrist, A. Muhammad, Blind swarms for coverage in 2-d, in Proceedings on Robotics: Systems and Science (2005)
13. C. Delfinado, H. Edelsbrunner, An incremental algorithms for Betti numbers of simplicial complexes on the 3-spheres. Comp. Aided Geom. Des. **12**(7), 771–784 (1995)
14. J. Eckhoff, Helly, Radon, and Caratheodory Type Theorems, Chap. 2.1 in *Handbook of Convex Geometry*, ed. by P.M. Gruber, J.M.Wills (North-Holland, Amsterdam, 1993), pp. 389–448
15. D. Estrin, D. Culler, K. Pister, G. Sukhatme, Connecting the physical world with pervasive networks. IEEE Pervasive Comput. **1**(1), 59–69 (2002)
16. S. Fekete, A. Kröller, D. Pfisterer, S. Fischer, Deterministic boundary recongnition and topology extraction for large sensor networks, in *Algorithmic Aspects of Large and Complex Networks* (2006)
17. D.W. Gage, Command control for many-robot systems, in Nineteenth Annual AUVS Technical Symposium, Huntsville, Alabama, USA (1992), pp. 22–24
18. A. Galstyan, B. Krishnamachari, K. Lerman, S. Pattem, Distributed online localization in sensor networks using a moving target (2003, preprint)
19. R. Ghrist, D. Lipsky, S. Poduri, G. Sukhatme, Node isolation in coordinate-free networks, in Proceedings of the Sixth International Workshop on Algorithmic Foundations of Robotics (2006)
20. M. Gromov, Hyperbolic groups, in *Essays in Group Theory*, MSRI Publ. 8 (Springer, New York, 1987)
21. A. Hatcher, *Algebraic Topology* (Cambridge University Press, Cambridge, 2002)
22. C.-F. Hsin, M. Liu, Network coverage using low duty-cycled sensors: random & coordinated sleep algorithms, in Proceedings of IPSN (2004)

23. C.-F. Huang, Y.-C. Tseng, The coverage problem in a wireless sensor network, in ACM International Workshop on Wireless Sensor Networks and Applications (2003), pp. 115–121
24. T. Kaczynski, K. Mischaikow, M. Mrozek, *Computational Homology, Applied Mathematical Sciences*, vol. 157 (Springer, New York, 2004)
25. H. Koskinen, On the coverage of a random sensor network in a bounded domain, in Proceedings of 16th ITC Specialist, Seminar (2004), pp. 11–18
26. S. Kumar, T. H. Lai, J. Balogh, On k-coverage in a mostly sleeping sensor network, in Proceedings of 10th International Conference on Mobile Computing and Networking (2004)
27. X.-Y. Li, P.-J. Wan, O. Frieder, Coverage in wireless ad-hoc sensor networks. IEEE Trans. Comput. **52**(6), 753–763 (2003)
28. B. Liu, D. Towsley, A study of the coverage of large-scale sensor networks, in IEEE International Conference on Mobile Ad-hoc and Sensor Systems (2004)
29. S. Meguerdichian, F. Koushanfar, M. Potkonjak, M. Srivastava, Coverage problems in wireless ad-hoc sensor network, in IEEE INFOCOM (2001), pp. 1380–1387
30. K. Mischaikow, M. Mrozek, J. Reiss, A. Szymczak, Construction of symbolic dynamics from experimental time series. Phys. Rev. Lett. **82**(6), 1144 (1999)
31. R. Moses, D. Krishnamurthy, R. Patterson, A self-localization methods for wireless sensor networks. EURASIP J. Appl. Signal Proc. (2002)
32. A. Muhammad, R. Ghrist, M. Egerstedt, Beyond graphs: higher dimensional structures for networked control systems. Control Sys. Mag. (2006)
33. Plex, version 2.1, Jan 2006, http://math.stanford.edu/comptop/programs/plex/
34. A. Rao, S. Ratnasamy, C. Papdimitriou, S. Shenker, I. Stoika, Geographic routing without location information, in Proceedings of ACM MOBICCOM (2003)
35. S. Simic, S. Sastry, Distributed environmental monitoring using random sensor networks. Lect. Notes Comp. Sci. **2634**, 582–592 (2003)
36. L. Vietoris, Uber den hoheren Zusammenhang kompakter Raume und eine Klasse von zusammenhangstreuen Abbildungen. Math. Ann. **97**, 454–472 (1927)
37. F. Xue, P.R. Kumar, The number of neighbors needed for connectivity of wireless networks. Wireless Netw. **10**(2), 169–181 (2004)
38. H. Zhang, J. Hou, Maintaining coverage and connectivity in large sensor networks, in International Workshop on Theoretical and Algorithmic Aspects of Sensor, Ad hoc Wireless and Peer-to-Peer Networks, Florida, Feb 2004
39. A. Zomorodian, G. Carlsson, Computing persistent homology, in Proceedings of 20th ACM-Symposium on Computing Geometry (2004), pp. 346–356

Chapter 5
Coverage Assessment and Target Tracking in 3D Domains

Recent advances in integrated electronic devices motivated the use of wireless sensor networks (WSNs) in many applications including domain surveillance and mobile target tracking, where a number of sensors are scattered within a sensitive region to detect the presence of intruders and forward related events to some analysis center(s). Obviously, sensor deployment should guarantee an optimal event detection rate and should reduce coverage holes. Most of the coverage control approaches proposed in the literature deal with two-dimensional zones and do not develop strategies to handle coverage in three-dimensional domains, which is becoming a requirement for many applications including water monitoring, indoor surveillance, and projectile tracking. This chapter proposes efficient techniques to detect coverage holes in a 3D domain using a finite set of sensors, repair the holes, and track hostile targets. To this end, we use the concepts of Voronoi tessellation, Vietoris complex, and retract by deformation. We show in particular that, through a set of iterative transformations of the Vietoris complex corresponding to the deployed sensors, the number of coverage holes can be computed with a low complexity. Mobility strategies are also proposed to repair holes by moving appropriately sensors toward the uncovered zones. The tracking objective is to set a non-uniform WSN coverage within the monitored domain to allow detecting the target(s) by the set of sensors. We show, in particular, how the proposed algorithms adapt to cope with obstacles. Simulation experiments are carried out to analyze the efficiency of the proposed models. To our knowledge, repairing and tracking is addressed for the first time in 3D spaces with different sensor coverage schemes.

This chapter has been reprinted with permission from "Coverage Assessment and Target Tracking in 3D Domains", Noureddine Boudriga, Mohamed Hamdi, S. S. Iyengar, Sensors, 2011.

DOI: 10.1007/978-1-4419-8420-3_5,

5.1 Introduction

One among the main WSN issues that should be addressed while dealing with target tracking and monitoring applications, in 3D environments with obstacles, is area coverage. This is because a sensor can detect the occurrence of events or the presence of hostile targets only if they are within its sensing range. Coverage reflects how well a zone is monitored or a system is tracked by sensors. Therefore, the WSN detection performance depends on how well the wireless sensors observe the physical space under control.

Several metrics have been provided in the literature to measure the quality of coverage. Among these metrics, one can mention the following: (a) the number of coverage holes; (b) the proportion of uncovered area with respect to the area under monitoring; and (c) the so-called average linear uncovered length (ALUL), which has been developed in 2D zones to estimate the average distance a mobile target can traverse before being detected by one sensor [1]. The ALUL can be used to assess the detection efficiency of the WSN in more general spaces. However, the major shortcoming of this approach is its heavy computational load making it nonconforming with the severe processing and energy limitations characterizing WSNs.

Obstacles in monitored 3D domains may complicate seriously the role of the monitoring sensors, increase their power consumption, and limit the coverage efficiency of the process providing coverage control [2, 3]. Procedures set up to implement coverage control and target tracking efficiency should be optimal. They should take into consideration the geographic nature of the monitored area and cope with the number and the shape of obstacles.

This chapter proposes a coverage assessment approach amenable to implement advanced target tracking functionalities. First, it provides a technique based on the concept of retraction by deformation applied to a special space, called the Rips complex, associated with the deployment of a set of sensors to develop a low-complexity algorithm for locating coverage holes. Second, it constructs a collaborative mechanism to repair coverage holes, assuming that the sensors have mobility capabilities. Third, the chapter builds on higher-order Voronoi diagrams to define an efficient scheme to coordinate tracking activities of single and multiple targets. To the best of our knowledge, this is the first time where retraction by deformation and higher-order Voronoi tessellations are used for hole assessment and target tracking in 3D domains with obstacles using sensors. The major contributions of this chapter are as follows:

- The definition proposed to distributively reduce the Rips complex associated with the sensors is general, in the sense that it applies to a large variety of sensor, detection techniques, monitored domains, and obstacles.
- The proposed cooperative coverage repairing approach considerably reduces the uncovered areas and provides efficient handling of obstacles with respect to existing methods. The detection and localization of holes is done with low complexity.

- We show that the higher-order Voronoi tessellations we utilize are useful for performing multiple tasks including activity scheduling and coordination. In addition, we show that local coverage information, when gathered using the Voronoi diagram, can be used to implement coverage-preserving mobility models.

The remaining part of this chapter is organized as follows: Sect. 5.2 describes the state of the art of coverage control in various areas in general and in 3D spaces in particular. Section 5.3 surveys the definition of the mathematical objects needed for coverage and tracking control, the Vietoris complex, and the Voronoi diagram and discusses the retraction by deformation. Section 5.4 discusses different schemes based on the Vietoris complex to detect and count the coverage holes in 3D domains, locate these holes, and repair them. It also defines a special procedure to reduce the complexity of the Vietoris complexes without modifying their topological properties. Section 5.5 sets up models for coverage assessment, sensor mobility, and target tracking. Section 5.6 analyzes the complexity of the algorithms constructed in this chapter and sets some extensions of our results to more general types of sensors. Section 5.7 develops simulation experiments to evaluate the performance of a monitoring system implementing our techniques. Section 5.8 concludes this chapter.

5.2 Related Work

Studies on coverage, holes, and boundary detection have been addressed using three main categories of techniques: geometric methods, statistical/probabilistic methods, and topological methods.

Studies using probabilistic approaches usually make assumptions on the probability distribution of the sensor deployment. Fekete et al. [4] assume uniformly randomly distributed sensors inside a geometric region for their boundary detection algorithm. Their approach hinges on the idea that the boundary nodes would have lower average degrees than that of the "interior" nodes and statistically provide a degree threshold to differentiate interior and boundary nodes. Kuo et al. [5] propose an error model for location estimation using probabilistic coverage, while Ren et al. [6] presents an analytical model based on probabilistic coverage to track moving objects in a densely covered sensor field. Most of probabilistic approaches have focused on the detection and tracking of objects in a sensor field. They did not address other related issues such as location of the holes, number of such holes, and repairing.

A number of literature has addressed the static or "blanket" coverage. Dynamic or "sweeping" coverage [7] has been also a common and challenging task with applications ranging from security to housekeeping. Two primary approaches to static coverage problems in the literature. The first uses computational geometry tools applied to exact node coordinates. Such approaches are very rigid with regard to inputs: one, for example, must know exact node coordinates and must know the geometry of the domain to determine the Delaunay complex. To alleviate the former requirement, many authors have turned to probabilistic tools. For example, in [8], the author assumes a randomly and uniformly distributed collection of nodes in a domain with a fixed geometry and proves expected area coverage. Other approaches give probabilistic or percolation results about coverage for randomly distributed nodes. The drawback of these methods is the fact that uniform distribution of nodes may not be always realistic.

More recently, the robotics community has explored how networked sensors and robots can interact and augment each other: (see e.g., [9] for more details). There are several new approaches to networks without localization that come from research works in ad hoc wireless networks that are not unrelated to coverage questions. One example is the routing algorithm of [10], which generally works in practice but is a heuristic method involving heat-flow relaxation. This work investigates the issues of maintaining coverage and connectivity by keeping minimum number of sensor nodes to operate in the active mode. The authors show that if the radio range is at least twice the sensing range, then complete coverage implies connectivity. A decentralized and localized density control algorithm, called OGDC, is devised to control and maintain coverage and connectivity. However, their approach requires knowledge of node location. The authors claim that this requirement can be relaxed to that each node knows its relative location to its neighbors. On the other hand, Hsin and Liu [11] give methods for localizing an entire network if localization of a certain portion is known. They address the problem of target tracking in face of partial sensing coverage by considering the effect of different random and coordinated scheduling schemes. In their coordinated coverage algorithm, a sensor might decide to sleep for some time after acknowledgments from its neighbor(s) that must be active. These decisions are not synchronized as individual sensors could negotiate with sponsors independently.

Since coverage verification is inherently a geometric problem, many research done in this area are based on computational geometry, and more precisely on the Voronoi Tessellation (and its dual, Delauney Triangulation). Motivated from the early success of the application of geometric techniques to cope with coverage problems (*Art Gallery Problem*), researchers have applied these techniques to ad hoc distributed sensor networks ([12–15]).

The most important drawback of these approaches is that they are too computationally expensive to be implemented in real-time contexts. Another severe limitation is the impact of localization uncertainty on the performance of these approaches. These claims are well documented in ([16]). In fact, to detect coverage holes, the locations of the sensors must be exactly known. Obviously, this cannot be always provided, especially when the sensing nodes are mobile. Moreover, equipping sensors with localization devices may considerably increase the deployment cost of the WSN and reduce its resources. In the following paragraphs, we summarize the methodologies, problems addressed, and results of some of the recent, notable studies in the area of detection and coverage in WSNs.

Meguerdichian et al. [13] study the problem of computing a path along which a target is least or most likely to be detected. They provide an optimal polynomial time algorithm that uses graph theoretic and computational geometric (Voronoi diagram) methods. They address the issues of maximal breach path and maximal support path and provide best- and worst-case coverage using computational geometry. Delaunay triangulation was used to find the best coverage path. In addition, deployment heuristics are provided to improve coverage. Since computational geometric methods require location information, the authors implement a location procedure prior to their coverage scheme. This procedure requires that a few of the deployed nodes (called beacons) must know their locations in advance (either from GPS or pre-deployment). Li et al. [14] uses local Delauney triangulation, relative neighborhood graph, and the Gabriel graph to find the path with the best-case coverage.

Huang et al. [15] study the problem of k-coverage. They propose solutions to the k-UC and k-NC (unit disks and non-unit disks) coverage problems which are modeled as decision problems whose goal is to determine whether reach location of a target sensing area is sufficiently covered. They present a polynomial time algorithm with a geometric approach that runs in $O(nd \log d)$ time.

Ghrist et al. [17] use topological methods to detect insufficient sensor coverage and holes. In their seminal work on using homological concepts for addressing hole detection and coverage, their algorithm detects holes with no knowledge of their location. Although the approaches by Ghrist et al. have many desirable properties, the assumption of a static network and the centralized scheme are not suitable for dynamic networks.

5.3 Mathematics for Coverage and Tracking

The objective of this section is to provide a mathematical model for accurately gauging the coverage degree of a monitored domain in the 3D space $\mathbb{R}^3$ and repairing the coverage holes. This model uses the Vietoris complex [6, 8].

The following assumptions will be used in the next sections: Let M be a bounded domain (or manifold) in $\mathbb{R}3$ with nonempty boundary ∂M. The boundary is assumed to be an orientable topological surface (i.e., a closed surface homeomorphic to some number of spheres and some number of connected sum of g tori, for $g \geq 1$, [18]). Let $\delta : \mathbb{R}^3 \times \mathbb{R}^3 \mapsto \mathbb{R}^+$ denoting the Euclidean distance. We denote by S a set of sensors deployed in $\mathbb{R}^3$ to monitor M, and by $|S|$ the number of these sensors. We will designate indifferently by $p \in S$ the sensor in S and its location (x_p, y_p, z_p) in $\mathbb{R}^3$. Let us notice, finally, that the sensors can be deployed inside M or outside it.

5.3.1 Voronoi Diagrams for Spherical Detection Sensors

Let us assume that the sensors in S have identical covered area represented by a ball with radius ρ. For every pair $p, q \in S$, we denote by $B(p, q)$ the plane, in $\mathbb{R}^3$ perpendicular to segment $[p, q]$ and passing by its middle point and by $H(p, q)$ the half space of $\mathbb{R}^3$ containing the p and delimited by $B(p, q)$. Thus, $B(p, q)$ and $H(p, q)$ are expressed as follows:

$$B(p,q) = \{x \in \mathbb{R}^3 | \delta(p,x) = \delta(q,x)\} \tag{5.1}$$

$$H(p,q) = \{x \in \mathbb{R}^3 | \delta(p,x) \leq \delta(q,x)\} \tag{5.2}$$

We also denote by $H_M(p,q)$ and $B_M(p,q)$ the intersection of $H(p,q)$ and $B(p,q)$ with M, respectively.

The Voronoi cell generated by $p \in S$ is nothing but the common area to the $(|S|-1)$ closed half-spaces containing p involving the other sensors. Therefore, the Voronoi cell generated by p is expressed by:

$$V_S(p) = \bigcap_{q \in S \backslash p} H(p,q) \tag{5.3}$$

The Voronoi cell of a sensor is convex and contractible. The common boundary of two Voronoi cells $V_S(p) \cap V_S(q)$ is included in $H(p,q)$. It can be a plane, a half-plane, an edge, a point, or an empty set. The Voronoi diagram associated with the set S of sensors deployed to monitor M is the unique subdivision defined in $\mathbb{R}^3$ by the Voronoi cells associated with all sensors. Thus, every cell of the subdivision contains the nearest neighbors defined in S for a sensor p. The Voronoi diagram of S is the set of point belonging to all the Voronoi cell. Hence, we have

$$V^D(S) = \bigcup_{p \in S} V_S(p) \tag{5.4}$$

In particular, the Voronoi diagram $V^D(S)$ has no vertices and no edges when the sensors are located at collinear points. In that case, the faces of the Voronoi diagram are parallel planes. In addition, one can notice that when $p \in S$ lies on the boundary of the convex hull of S, then the Voronoi cell of p is unbounded in $\mathbb{R}^3$.

Since in this chapter, we are rather interested in partitioning a domain M into cells according to k-nearest neighbors in S, for a given integer $1 \leq k \leq n-1$, we turn now to the definition of the high-order Voronoi diagrams, as they are useful concepts to define these sets and support target tracking. An order k Voronoi diagram is defined as follows: Let $T \subset S$ containing k sensors, the T-generated cell is defined by:

$$V(T) = \{x \in \mathbb{R}^3 | \forall p \in T \forall q \in S - T : \delta(x,p) \leq \delta(x,q)\} \tag{5.5}$$

The order k Voronoi diagram is given by:

$$V_k^D(S) = \bigcup_{\substack{T \subseteq S \\ |T|=k}} V(T) \tag{5.6}$$

One can easily see that the order 1 Voronoi diagram $V_1^D(S)$ is just $V^D S$, that $V(T)$ can be empty, and that $V^D(S)$ induces a partition on the domain M into bounded components.

5.3.2 Vietoris–Rips Complexes

We consider a set of points $S = \{v_1, \ldots, v_n\}$ corresponding to the locations of a set of sensor nodes in a 3D space. For brevity, $(v_i)_{1 \leq i \leq n}$ will be simply referring indifferently to as sensor nodes and points. We suppose that each sensor is capable of covering a disk of radius r_c and communicate with the other sensors within a distance $r_b \leq \sqrt{3}r_c$. The total region covered by the sensor network can be represented by:

$$\Gamma(S) = \bigcup_{v_i \in S} \Gamma_{v_i, r_c} \tag{5.7}$$

where:

$$\Gamma_{v_i, r_c} = \{x \in \mathbb{R}^3 |\ ||x - v_i|| \leq r_c\}$$

A k-simplex (or a simplex of dimension k) σ is an unordered set $\sigma = \{v_0, v_1, \ldots, v_k\} \subseteq S$, where $v_i \neq v_j$ and $\delta(v_i, v_j)$, for all $i \neq j$. A face of the k-simplex σ is a $(k - 1)$-simplex formed by k elements (or vertices) of σ. Clearly, any k-simplex has exactly $k + 1$ faces. The collection of all k-simplices of S is called the abstract associated with $\Gamma(S)$. In fact, an abstract simplicial complex X is a finite collection of simplices which is closed with respect to the inclusion of faces; meaning that if $\sigma \in X$, then all faces of σ are also in X. It is noteworthy that a simplicial complex is a generalization of a graph; that is, the connectivity graph is nothing but the set of 1-simplices of the simplicial complex associated with a set V of points in the 3D space.

Now, let us discuss the definition of the Vietoris–Rips complex. This complex captures the features related to connectivity and coverage of WSNs.

Definition 5.1 (*Vietoris–Rips complex*) Let S be a set of points in a 3D space and a given radius ϵ. The Vietoris–Rips complex of S, denoted by $R_\epsilon(S)$, is the simplicial complex whose k-simplices correspond to unordered $(k + 1)$-tuples of points in S which are pairwise within Euclidean distance ϵ of each other.

A subset of $k + 1$ points in S determines a k-simplex of for the Vietoris–Rips complex if, and only if, each of these points lies within the intersection of the balls of radius ϵ centered at the other k points.

The reader, however, may wonder whether such topological structure can be computed in practice by tiny motes equipped with radio devices and limited storage capabilities. To answer this question, we propose a simple mechanism allowing a fully distributed construction of the Vietoris–Rips complex. Through a 3-step broadcast of connectivity information, each sensor node can be aware of what simplices it belongs to, and what other simplices its neighbors belong to. To this end, we assume that every sensor node has a unique identifier (typically a layer-2 address) and has enough space to maintain a table of identifiers. The protocol performs as follows:

1. Initialization: Every sensor v_i broadcasts its identity to its neighbors. Upon receipt of the message, each sensors builds the list, denoted by Σ_0^i, of 0-simplices formed by its neighbors.
2. Edge construction: Sensor v_i appends its identity to the vertices in Σ_0^i to construct the list, say Σ_1^i, of all 1-simplices it belongs to. It also determines the number n_i of its neighbors. Then, it informs its neighbors about the 1-simplices it built.
3. Simplicial iteration: On receiving the information from its neighbors, sensor v_i starts building the lists Σ_j^i , $2 \leq j \leq n_i$, by simply adding appropriately the structures it has received to the ones it has already constructed.

An informal explanation of the construction algorithm is as follows. Simplices of higher dimension are constructed iteratively. In the first iteration, the 2-simplices are constructed by applying the following rule:

$$\langle v_i, v_j\rangle \in \Sigma_1^i \wedge \langle v_i, v_k\rangle \in \Sigma_1^i \wedge \langle v_j, v_k\rangle \in \Sigma_1^j \rightarrow \langle v_i, v_j, v_k \in \Sigma_2^i$$

for every i, j, and k, provided that $i \neq j, i \neq k, j \neq k$. The rules used for the following iterations are similar.

5.3.3 Homotopy and Retraction

Let X and Y be two topological spaces and $f, g : X \mapsto Y$ be two maps (or continuous functions). We say that f and g are homotopic if there is a map $F : [0, 1]times[0, 1] \mapsto X$ such that

$$F(x, 0) = f(x) \wedge F(x, 1) = g(x) \qquad \forall x, y \in X$$

Let $x_0 \in X$ be a given base point of X. A loop based on x_0 is a map $\alpha : [0, 1] \mapsto X$, such that $x0 = \alpha(0) = \alpha(1)$. An equivalence relation on the set of all loops based at x_0 can be defined by stating that loops α_1 and α_2 are equivalent if they are homotopic with respect to x_0, meaning that there exists a homotopy F between α_1 and α_2 such that

$$F(0, t) = F(1, t) = x_0 \qquad \forall t \in [0, 1]$$

We denote the equivalence class of a loop $\alpha : [0, 1] \mapsto X$ based at x_0 by $[\alpha]$ and call it the based homotopy class of the loop α. The set of equivalence classes of loops based at x_0 is denoted by $\pi_1(X, x_0)$ and is called the fundamental group. It can be equipped with a multiplication defined by $[\alpha_1] \times [\alpha_2] = [\alpha_1 \cdot \alpha_2]$, for all loops $[\alpha_1]$ and $[\alpha_2]$ based at x_0, where $\alpha_1 \cdot \alpha_2$ is the loop obtained by attaching α_1 to α_2. A second group of homotopy, denoted by $\pi_2(X, x_0)$, can be defined as the set of homotopy equivalence classes of applications $\beta : [0, 1]^2 \mapsto X$, based at x_0. It is an Abelian group, [19].

On the other hand, a map $f : X \mapsto Y$ is called a homotopy equivalence if there is a map $g : Y \mapsto X$ such that $f \circ g$ is homotopic to the identity function in X and

$g \circ f$ is homotopy to the identity function Y. Thus, one can say that two spaces are homotopy equivalent if they have "the same shape."

A deformation retraction of a space X onto a subspace $A \subseteq X$ is a map $f : X \times [0, 1] \mapsto X$ such that:

$$f(x, 0) = x \wedge f(x, 1) \in A \wedge f(a, t) = a \qquad \forall x \in X \forall a \in A \forall t \in [0, 1]$$

In other words, the subset A is a retraction by deformation of the space X if, starting from the original space X at time 0, we can continuously deform X until it becomes the subspace A at time 1 and deformation is performed without ever moving the subspace A in the process. It is obvious that if A is a retraction by deformation of X, then X and A are homotopically equivalent.

Finally, let K be complex, a retraction filtration of K is a nested finite sequence of subcomplexes K_i,

$$K_0 \subseteq K_1 \subseteq \ldots \subseteq K_n = K$$

such that, for all $k \geq 0$, K_k is a retract by deformation of K_{k+1}: Thus, it can be shown easily that K_0 and K_n have the same type of homotopy and the same homotopy group.

Let $T = \{p_1, \ldots, p_k\}$ be a simplex and $T_1 = \{p_2, \ldots, p_k\}$ be one of its faces. Then, $A = T - (T_1 - \partial T_1)$ be the part of the boundary of T that is not internal to T_1. Then, A is a deformation retract of T.

Let $R(S)$ be the Rips complex associated with S, repeating the process of retraction of simplexes that are on the boundary of $R(S)$, with faces external to $R(S)$, would lead to a filtration of $R(S)$; say K_k, $0 \leq k \leq n$, such that, for all $k \geq 0$, K_k is a retract by deformation of K_{k+1} and K_{k+1} is obtained from K_k by adding one simplex, external to K_k and belonging to $R(S)$. The object K_0 has no simplex with external face that is retractible.

5.4 Coverage Hole Management of Spherical Sensors

In this section, we propose a novel distributed technique to count the coverage holes of WSN using the retraction theory of spaces. In particular, we show that the Vietoris–Rips complex associated with the WSN can be reduced to a simpler space that is tightly related to the number of holes.

In the following, let $D \subseteq \mathbb{R}^3$ be a compact domain in the 3D space $\mathbb{R}^3$ and ∂D be its boundary. We consider that D contains no obstacles. We also consider that a collection $S = \{v_1, \ldots, v_n\}$ is deployed over domain D and that the sensors are equipped with local communication and sensing capabilities. In fact, each sensor is capable of communicating directly with other sensors in its proximity (within a given distance r_b) and has a limited sensing range ϵ.

5.4.1 Reducing the Vietoris–Rips Complex

We assume, in this section, a complete absence of localization capabilities and metric information, in the sense that the sensors in the network can determine neither distance nor direction. Under these assumptions, we are interested in designing distributed algorithms for coverage assessment and hole detection.

To this end, we need first to introduce a special procedure, called retract, that reduces the size of the Vietoris–Rips complex while keeping its type of homotopy. Repeating this procedure several times will eliminate all the 3 cells of the Vietoris–Rips complex.

Let $R_\epsilon(S)$ be the Vietoris–Rips complex. Let $\{v_0, \ldots, v_3\}$ be a 3-simplex in $R_\epsilon(S)$ such that one of the 2-cell $\{v_0, v_1, v_2\}$ does not belong to another 3-simplex in $R_\epsilon(S)$. If such a situation does not exist, then one can easily deduce that $R_\epsilon(S)$ has no 3 cells. Let X_1 and A_1 be the set of points $x \in R_\epsilon(S)$ belonging to simplex $\{v_0, v_1, v2\}$ and the subset of X_1 generated by the other two faces, respectively. Then, it is easy to construct a map $h_1 : X_1 \times [0, 1] \mapsto X_1$ such that:

$$h_1(x, 0) = x \wedge h_1(x, 1) \in A_1 \wedge h_1(a, t) = a \qquad \forall x \in X \forall a \in A_1 \forall t \in [0, 1]$$

Map h_1 can be easily extended to a map

$$\text{Retract} : R_\epsilon(S) \times [0, 1] \mapsto R_\epsilon(S)$$

such that:

$$\text{Retract}(x, 0) = x \wedge \text{Retract}(x, 1) \in A \wedge \text{Retract}(a, t) = a$$
$$\forall x \in X \forall a \in A \forall t \in [0, 1]$$

where A is $\big(R_\epsilon(S) - X_1\big) \cup A_1$.

Repeating the map Retract several times will lead to eliminating all the 3-simplices in $R_\epsilon(S)$. The map Retract can also be reapplied several times to delete all 2-simplices and 1-simplices that a free face. The resulting space, say R_ϵ^{red}.

Proposition 5.1 *Let S be a set of sensors. If $R_\epsilon(S)$ is path-connected, then $R_\epsilon^{\text{red}}(S)$ satisfies the following properties:*

1. *R_ϵ^{red} is homotopy equivalent to $R_\epsilon(S)$.*
2. *the number of holes delimited by $R_\epsilon^{\text{red}}(S)$ is equal to the number of holes of the vietoris space $R_\epsilon(S)$*

Proof Applying the map Retract several times helps creating a retraction filtration of $R_\epsilon(S)$ such that:

$$R_\epsilon^{\text{red}}(S) = K_0 \subseteq K_1 \subseteq \ldots \subseteq K_n = R_\epsilon(S)$$

where n is number of 3-simplices in $R_\epsilon(S)$. Since, for every i, K_i is homotopy equivalent to K_{i+1}, we can deduce that $R_\epsilon^{\text{red}}(S)$ is homotopy equivalent to $R_\epsilon(S)$.

The second statement of the theorem can be deduced from the following features:

- a holes is a path-connected component that is surrounded by the delimiting space ($R_\epsilon^{\text{red}}(S)$ and $R_\epsilon(S)$).
- Retracting a 3-simplex in $R_\epsilon(S)$ may enlarge a hole but does not eliminate it.
- The retraction process does not create holes since it operates on the simplices that have free faces.

□

5.4.2 Counting and Locating Coverage Holes

To count and locate holes, we set up a 3-step algorithm. In the first step, we construct the external boundary of $R_\epsilon(S)$. This is the subset of S containing all the nodes occurring on free faces and facing the boundary ∂D of the domain. In the second step, we define an algorithm that detects holes by progressively transforming the external boundary by retracting all its external simplices. In the third step, the following process is repeated: one external 2-simplex is deflated, the Retract map is applied several times to reduce appearing simplices with free faces, and the external boundary is updated. The number of iterations of this process gives the number of coverage holes.

5.4.2.1 Constructing the Boundary of $R_\epsilon(S)$

Let us assume that the boundary ∂D of the domain D under monitoring can be seen (or detected) by the sensors in S and that the nodes in S broadcast periodically their unique ID numbers. The construction is based on the three following actions:

- Every sensor node detecting a boundary component of D or finding itself on an external facet sends this information to its neighbors.
- The information related to boundary detection, when received by sensors, should be put together to form the external boundary of $R_\epsilon(S)$, by simply allowing every sensor node to know which neighbor is on the external boundary.
- The nodes broadcast information related the external boundary of $R_\epsilon(S)$ so that every node on the boundary can have a precise picture of the boundary.

5.4.2.2 Counting Coverage Holes

Counting the coverage holes can be set up by an algorithm that repeats iteratively the following major procedures:

- Boundary retraction: Let C_n be a n-simplex on the boundary of $R_\epsilon(S)$ and C_{n-1} be one of its external faces, then C_n can be retracted using the procedure Retract and the boundary is updated by adding a new node (the one in $C_n - C_{n-1}$), if $n \leq 2$, or by deleting the node occurring in C_{n-1}, if $n = 1$.
- Boundary deflation: When all the simplices on the boundary have been retracted, a preselected node in S (in charge of the counter) selects one of the nodes of the new external boundary, withdraws it from the boundary, and increments the counter.

5.4.2.3 Locating Coverage Holes

It is worth noticing that when a deflation of a 2-simplex on the boundary $R_\epsilon(S)$ is applied after retraction is complete, a hole is reduced from the coverage zone. This because the selected node, for deflation, is observing the hole, since it is one of the nearest nodes surrounding the reduced hole. Thus, this node can start the construction of the boundary of the reduced hole by determining the list of the nodes surrounding immediately the hole.

One can conclude, therefore, that any time a deflation is operated, a hole can be located by simply constructing its boundary using the nearest nodes to that hole.

5.4.3 Repairing Coverage Holes

Let us here assume that the 3D domain D under monitoring has no obstacles and let us denote by χ ($\chi = 4\pi\epsilon^3/3$) the volume of the area covered by a sensor and by $\mathrm{Vol}(D)$ the volume of D. One can state that the number $|S|$ of sensors in S should be higher than the number $N_0 = \mathrm{Vol}(D)/\chi$ to be able to guarantee full coverage of D, at least after hole detection and coverage optimization. Therefore, we will assume in the sequel that this condition is satisfied. Finally, we assume that the sensors are able to move and detect the external boundary of D, when they are close to it, like in the above section.

Repairing holes aims at extending the coverage by eliminating the holes, or at least by shrinking considerably their size. An algorithm can be defined to this purpose. It can be built based on the following general rules:

- A node detecting the external boundary ∂M should keep seeing the boundary when it moves.
- A node on the external boundary of $R_\epsilon(S)$ should move toward the uncovered area, when it does not see the boundary.
- When two neighbor nodes on the external boundary of $R_\epsilon(S)$ are separated by a distance higher than a predefined threshold, say θ_1, and one of them is not seeing the boundary of D, then the sensor unable to see the boundary asks its successor (i.e., a neighbor involved in the retraction of the simplex containing this sensor) to move toward the external boundary.

- A node seeing the boundary should inform its neighbors so that they can move accordingly.
- When the distance between a sensor s and its neighbors on the boundary of a hole is lower than a predefined value, say θ_2, then s should move in the opposite direction of the hole, while the other sensors should move toward the hole so that when they see each other, s can withdraw itself from the minimal surface after informing its neighbors.
- A node on the external boundary finding itself unable to move informs its successor to move toward its direction.

5.5 Target Tracking in 3D Domains

In this section, we use 3D Voronoi diagrams to optimize sensor coverage and target tracking performance. We first propose a strategy to measure the uncovered zones of the monitored region. Then, we develop two mobility models that provide target tracking using order k Voronoi diagrams and optimize the coverage ratio of a zone using Voronoi cells. Finally, we extend these models to multiple target tracking. We assume in this section that the sensors have spherical coverage. The vector-guided case can be addressed using similar techniques.

5.5.1 Measuring Uncovered Areas

Assume that a location x within the surveillance area is not covered by any sensor. Let $\mathcal{L}(x, \theta)$ define the Linear uncovered length (LUL) at location x with direction θ. This is the undetected path length of a target traveling from location x with direction $\theta = (\theta_1, \theta_2)$, for $0 \leq \theta_1 \leq 2\pi$, $-\pi/2 \leq \theta \leq \pi/2$).

The ALUL, denoted by ALUL(x), introduced in [20, 21], for the 2D space, gives an approximation of the average distance that can be made by a target, moving in 3D space, before being detected by the sensor network. The ALUL function can be defined by the following formula:

$$\mathrm{ALUL}(x) = \begin{cases} 0, & \text{if } x \text{ is covered.} \\ \frac{1}{(2\pi)^2} \int_{-\frac{\pi}{2}}^{\frac{\pi}{2}} \int_{0}^{2\pi} \mathcal{L}(x, \theta_1, \theta_2) \mathrm{d}\theta_1 \mathrm{d}\theta_2, & \text{otherwise.} \end{cases}$$

More generally, when A is a subregion of the 3D domain under supervision, the ALUL related to A, ALUL(A), that a target can travel within A without been detected by a sensor is given by the expression:

$$\mathrm{ALUL}(A) \equiv \frac{\int_{x \in A} \mathrm{ALUL}(x) \mathrm{d}x}{||A||} \tag{5.8}$$

where $||A||$ is the volume of A.

The ALUL metric was developed to deal with a static deployment, which is not the case of our study. When a mobility model is implemented, the topology of the WSN is no longer static. To overcome this, we extend this notion so as to support sensor node mobility. The ALUL should also vary according to time and should use a function, denoted by the $\mathcal{L}(x, \theta, t)$, that defines the LUL at location x with direction θ, at time t. Based on this reasoning, we define the metric $\text{ALUL}_m(x, t)$ representing the ALUL in a location x at time t.

Due to sensor node mobility, the ALUL, over time, in a point x will be expressed by:

$$\text{ALUL}_m(x) = \int_0^\infty \text{ALUL}_m(x, t)\mathrm{d}t \tag{5.9}$$

Finally, $\text{ALUL}_m(A)$ can be computed by Eq. (5.8) by replacing $\text{ALUL}(x)$ by $\text{ALUL}_m(x)$.

From the performance evaluation perspective, two important points should be highlighted:

- $\text{ALUL}_m(A, t)$ gives information about the coverage-preserving capabilities of the mobility model. It can be used to state whether the steady state is rapidly reached and whether the mobility model affect the detection performance of the sensor network.
- $\text{ALUL}_m(A)$ provides information about the long-term behavior of the mobility model. It can be used to evaluate the impact of mobility on the possibility for a target to be undetected within the monitored region.

5.5.2 Mobility Models for Target Tracking

In this section, we show how the Voronoi cells can be used to implement target tracking using a sensor mobility model. In fact, we define two mobility models:

- The first model is called k-mobility model. Sensor nodes in this model move toward the regions where the hostile target is supposed to be and collaborate to keep the target controlled by k sensors all the time. To this end, the order k Voronoi diagrams are used and maintained all the time.
- The second model is called simplified model. It relies on estimating the uncovered zones within the Voronoi cells, using the ALUL metrics and moving sensor nodes toward the "uncovered zones."

While the first model is triggered by the occurrence of targets, the second model aims at adapting the covered area so that the targets can be detected with higher probabilities. Obviously, the k-mobility model is more energy consuming than the second since it encompasses the prediction of the target position and requires tracking using k sensor nodes. Therefore, we suppose that the second model can be used when energy resources become scarce. The performance of both models will be assessed in Sect. 5.7. Moreover, one can notice that the prediction function we are using is tightly related to the coverage of the zones where the targets are expected and that the mobility models assume that nearest sensor nodes can move to these zones while reducing the coverage of other zones where targets are not expected. In fact, the greater is the number of target detection signals, the better is the prediction precision to command sensor movements.

5.5.2.1 The k- Mobility Model

In the following, we distinguish two cases: (a) a target crossing a k-covered area and (b) a target crossing non-k-covered zone.

5.5.2.2 For a Target Crossing a k-covered Zone

The mobility algorithm is triggered upon the detection of a target presence. Every detecting sensor sends its detection signal to the relevant intermediate sensor (called IS). The latter collects all detection signals, verifies their integrity, deduces the current zone where the target might be, estimates the positions of the target in the next of time slot, and commands k sensors to move to monitor the new zone to ensure tracking continuity.

Typically, the selected zone of target presence is taken among other zones (when more than k sensors detect the target presence). These zones are ordered according to the probability of presence of the target. The zone selected is the one presenting the highest probability among those which are k-covered.

The mobility algorithm is defined through five steps:

1. Assume that k' sensors detect the target ($k' > k$). The k' sensors si, $1 \leq i \leq k'$, send their detection data d_i to an intermediate node under the form:

$$d_i = (r_{t,i}, \theta_{t,i}, \tau_{t,i}, s_i)$$

   where $r_{t,i} = \delta(x_i, z_{t,i})$ is the Euclidean distance separating s_i from the position $z_{t,i}$ of the target as seen by s_i, $\theta_{t,i} = (\alpha_{t,i}, \beta_{t,i})$ is the direction of the vector $\overrightarrow{z_{t,i} - x_i}$, and $\tau_{t,i}$ is the detection instant.
2. In the case where detection signals are sent to different intermediate nodes, the intermediate nodes coordinate to gather all signals (or at least k of them) at a unique node IS, which verifies first the authentication of the messages.
3. IS constructs:

   - The zone of target presence $Z^{\tau}_{t,i}$ for each sensor based the errors made for the values reported. This zone is delimited by the following eight points:

$$(r_{t,i} \pm \Delta r, (\alpha_{t,i} \pm \Delta\alpha, \beta_{t,i} \pm \Delta\beta))$$

     as defined by the estimated detection errors.

- The most likely target presence zone $Z^{\tau}(t)$. Several strategies can be used for this including selecting the largest intersection of k zones of the form $Z^{\tau}_{t,i}$. It can also be the largest union of k zones. Let T be the set of k sensors involved in the definition of $Z^{\tau}(t)$.

Then, IS computes the order k Voronoi cell $V^S(T)$. Obviously, it contains $Z^{\tau}(t)$.

4. IS estimates the zone $Z^{\tau,+}(t)$, where target z_t is likely to be in the next time slot. Several strategies can be used for this estimation including extrapolation of older positions or some information related to target direction and speed. It also estimates the most likely new position of z_t.
5. IS selects k sensors based on a specific criteria and order them to move toward $Z^{\tau,+}(t)$ to increase its coverage. If no criteria is used, then the order goes to the sensors in T. A criteria can simply to reduce sensor movement.

When a criteria is applied for the selection of k sensors to cover the new position, some of the selected sensors (say k'' sensors) may belong to T and the others (say $k - k''$) have to be added among the neighbors of T. This situation is addressed in the following section.

5.5.2.3 For a Target Crossing a Non-k-covered Zone

In this case, only k' ($k' \leq k$) detection signals are received by the intermediate sensor IS, which should proceed at the construction of the probable current zone of presence of the target the way the preceding algorithms does. Then, it starts the selection of the remaining $(k - k')$ required signals. Then, it orders the movement of the k sensor providing k monitoring to the target. For this purpose, IS executes the following steps:

(1) IS computes the most likely zone of target presence; let z_t using the k' reports from k' sensors denoted by $s_1, \ldots, s_{k'}$.

(2) For each $i \leq k'$, IS selects the nearest k sensors to s_i. It computes the related k-Voronoi cell $V_i^{(k)}$ and deduces the intersection $z_t \cap V_i^{(k)}$

(3) For each $i \leq k'$, IS gets the number of sensors k_i'', $0 \leq k_i'' < k$ that have sent detection signals to IS.

(4) IS classifies the k-Voronoi cells according to the value of k_i''. The greater k_i'' is, the most important is the probability of presence of the target in $V_i^{(k)}$. A small value of k_j'' induces that the target is going in or out the cell $V_i^{(k)}$.

(5) IS selects the nearest k sensors involved in $\partial V_i^{(k)}$, where $k_i'' = \max_{j \leq k'} k_j$, and guides the $(k-k'')$ added sensors (among the nearest sensors to s_i) to move toward $\partial V_i^{(k)}$. For that, it sends them a mobility instruction including the probability of presence of the target. A mobility instruction is defined by the 3-tuple.

$$(r_i, \alpha_i, \pi_i)$$

where $r_i \geq \delta(s_i, p)$ such that $\forall q \in \partial V_i^{(k)} : \delta(s_i, p) \geq \delta(s_i, q)$ and $i = \arg\max \widehat{xs_i y}$ where $x, y \in v_i$ and v_i is the set of the vertices of the boundary ∂V_i, $\pi_i = k''/k$ is the probability of presence of the target in $\delta V_i^{(k)}$.

To enhance coverage while keeping more mobility freedom, we implement a group mobility model in which ground sensors move in groups in order to preserve the k-coverage. To this purpose, for each mobility step, the sensors define randomly groups of k members for each, the latter are not required to be the nearest neighbors. Each group chooses randomly a head which chooses the first mobility step. The remaining members of the group take into account this choice to determine their next mobility step. By this way, every sensor's mobility will depend on the integrating group. Furthermore, a sensor may move from one group to another in each mobility step. This model enables the definition of overlapping k-Voronoi groups which increases the guarantee of having a k-coverage.

5.5.2.4 Simplified Mobility Model

We propose hereafter a mobility model which is based on the use of simple Voronoi diagram to identify and reduce coverage holes.

This model can serve to implement a mobility strategy where a sensor node looks for one or more neighbors that are at least 2ϕ-distant from it. If such nodes exist, the sensor node moves toward the most distant neighbor, denoted by n_f, with a distance $\frac{\delta(s_i, n_f) - 2\phi}{2}$.

The following result extends this strategy to the case where the monitored region is required to be k-covered using the simplified algorithm. It uses a set, denoted by $X(s_i, V(S))$, which defined the set of intersection points expressed as follows:

$$X(s_i, V^D(S)) = \mathfrak{D}\big(V^D(S \setminus \{s_i\})\big) \cap \Gamma(s_i, R_{s_i}) \tag{5.10}$$

where $\mathfrak{D}$, for a region $R \subseteq \mathbb{R}^3$, denotes the boundary of R.

For the sake of parsimony, we do not provide proofs for these corollaries in this chapter.

Lemma 5.1 *For s_i in S, if $|N\big(s_i, V^D(S)\big)| < k$, where $|\cdot|$ denotes set cardinality, then $V^D(s_i)$ is not k-covered. For s_i in S, if $|X\big(s_l, V^D(S)\big)| < k$, then $V^D(s_i)$ is not k-covered.*

This lemma shows how simple Voronoi diagrams can be used to detect the coverage holes based on the distance between the sensor node and the edges of its Voronoi cell. It is based on the concept that the Voronoi tessellation is a partition of the points belonging to the monitored area according to their proximity to the sensor nodes. In other terms, if a point is not detected by the sensor node located at the generator of the Voronoi cell it belongs to, it cannot be detected by any other sensor node. If a sensor detects that the distance to one among the edges of its Voronoi edges is more than its coverage range, it has to move toward this edge to cover the corresponding hole. The uncovered can therefore be gradually reduced using this distributed strategy. However, a sensor node can detect that more than one of its Voronoi neighbors do not fulfill the condition of the lemma; it will therefore move toward the most distant neighbor.

The major advantage of this strategy, with respect to the advanced strategy, is that it relies on simple Voronoi diagrams to deal with k-coverage, while the advanced model proposed in the previous section is based on order k Voronoi tessellations which are more complex to build.

A more accurate comparison between the two models will be carried out in Sect. 5.7.

5.5.3 Multitarget Tracking

The two tracking models presented in the above can be extended to the tracking of multiple targets. To describe the extension, let us assume, for the sake of clarity, only two targets are detected by sensors in S. Let z_t and $z_{t'}$ be the reported positions.

The extension of the simplified model considers two cases:

- Only one node has detected the presence of the two targets: In that case, the sensor keeps monitoring one of the targets and invites the nearest neighbor to the second target to monitor the second and provides it with relevant information it collects.
- More than one node have detected the targets: In that case, two sensors among those that have detected the targets are selected to keep monitoring the targets independently.

On the other hand, the k-mobility model extends in following way: If d nodes detect the targets, these sensors are divided into two subsets, each in charge of monitoring one target; then, the subsets are extended so that any of them contains k sensors.

5.6 Complexity Analysis of Coverage Management and Tracking

5.6.1 Complexity

In this section, we analyze the complexity of the different algorithms we have developed in the previous sections to detect and locate holes or to repair coverage holes. Our approach to estimate the complexity can be based on the following metrics:

- The number of messages exchanged between the sensors during the execution of the algorithm.
- The number of additions and deletions of simplices to the Vietoris complex.
- The number of sensor movements made during the execution of an algorithm.

Some other operations can be added for a more accurate estimation of complexity. These metrics may include, for example, the number of storing operations made at the node level to update the related data structures. The messages exchanged during the execution of an algorithm can be of different types. In particular, they can be sent to a neighbor to tell it to change its status from internal (to the Vietoris complex) to external (i.e., on the boundary of Vietoris complex). They also can be used to construct the initial boundary of Rips complex, or used to reduce the external boundary. They also can be sent after the retraction or the deflation of a simplex, or they are sent by a leader node the command a coordinated movement of sensors.

For the sake of clarity, we will focus on the complexity on the detection and counting of coverage hole. In this case, let n be the number of sensors in S, and e be the number of 1-simplices, f the number of 2-simplices, and t the number 3-simplices in RIPS complex of S. Let also p the number of vertices at the initial boundary of the Rips complex.

The number of messages sent during the execution of the algorithm should be lower or equal to the number of messages exchanged if all the polyhedra (external and internal) have been retracted first and that after deflation, all the facets have been retracted. In that case, one can state that the number N_1 of messages sent is given by:

$$N = p + (f - p) + (e - (f - p)) = p + e \leq |S| + e$$

where p is the number of external vertices. This result can be deduced from the preceding and the fact that $p \leq |S|$.

Now let us assume, without loss of generality, that the deployment of sensors (initial and current) guarantees that every node in the Rips complex of S has at most v neighbors (v is a fixed value coping with the volume of the area to monitor and the radius of coverage). Then, one can conclude that e is smaller than $v \times |S|$, and we deduce that:

$$N \leq n + v \times |S| \leq (v + 1)n$$

Let us notice that the assumption is not mandatory and a direct proof can be given. This shows that the algorithm to detect and count the coverage holes has a linear complexity.

5.6.2 Extending Results

The results presented in the previous sections can be extended in two dimensions: the type of the sensors and the occurrence of obstacles in the domain under monitoring.

Coping with obstacles. The algorithms developed in the preceding sections can be adapted to the occurrence of obstacles. Obstacles in monitored 3D areas may complicate seriously the role of monitoring sensors, increase their power consumption, and limit the coverage efficiency. Two particular objects have to be modified in our algorithms. First, the coverage holes that have to be counted should not contain obstacles. One can assume for this that the sensors are able to recognize an obstacle. Second, the mobility model used to increase coverage or to provide tracking should consider moving the sensor vertically as an alternative.

Coping with semispherical. The algorithms developed in the preceding sections can be extended to semispherical sensors (sensor having a semispherical covered area). It is worth noticing, at this point, that this type is sufficiently general to represent various sensor-based applications. In particular, the model can be used to represent fire and smoking-based sensors or camera-based sensors. To cope with semispherical sensors, one can notice that the concepts of Vietoris–Rips and Voronoi diagram can be extended, so that coverage holes can be handled in a similar way. However, when repairing a hole, the mobility model of the sensor should include rotating a sensor to increase the coverage of a specific area by the sensor.

It is worth noticing that the additions made to the developed algorithms do not modify significantly the complexity of the algorithms. In particular, the complexity of the hole counter remains linear (as shown in the simulation discussed in the following section).

5.7 Experimental Results

In this section, we carry out a set of experiments to prove the efficiency of the proposed techniques. We first address the coverage hole problem by evaluating the performance of the higher-order Voronoi-based strategy for coverage optimization. To this purpose, we define a metric representing the ratio of uncovered area with respect to the total area of the monitored region. Second, we assess the target tracking approach by estimating the maximum linear distance that can be made by a hostile target without being detected. Finally, we evaluate the complexity of our coverage control and mobility techniques. We use the number of transmitted messages as a main criterion to estimate this complexity since data transmission consumes much more power than computational steps in WSNs.

5.7.1 Coverage Control and Hole Reduction

The first experiment aims at evaluating the hole reduction strategy based on three-dimensional Voronoi tessellations with spherical coverage. We define the following metric to evaluate the performance of hole reduction.

$$\mu = \frac{\text{Sum\_of\_hole\_volumes}}{\text{Total\_volume\_of\_the\_monitored\_area}} \tag{5.11}$$

Figure 5.1 shows the evolution of μ according to the number of iterations of the coverage hole reduction algorithm. We compared the Voronoi-based hole reduction

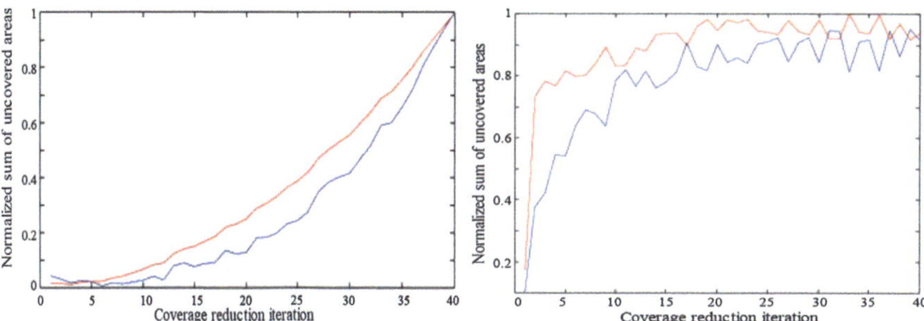

Fig. 5.1 Evolution of the uncovered area proportion according to time. The *left* plot corresponds to the spherical coverage, while the *right* one is related to the semispherical coverage

strategy with the homotopy-based strategy proposed in Sect. 5.4. It can be noticed that the increase in terms of normalized uncovered proportion is about 15 % when the number of iteration is low. In addition, when the number of iteration exceeds 30, both approaches perform well since the normalized sum of uncovered areas becomes higher than 90 %.

A similar experiment is conducted for vector-guided sensors, assuming that at every step of the iteration, the mobility is provided along with an orientation of the vector to achieve better coverage. Figure 5.1 shows the evolution of μ according to the number of iterations of the coverage hole reduction algorithm and compares it to the Voronoi-based hole reduction strategy with the homotopy-based strategy. One can conclude that while the homotopy-based approach is less complex, since linear for detection and localization, the Voronoi-based method reaches better results. In addition, a comparison between the results obtained for spherical sensors and semispherical sensors shows the following:

- The approach performs better with spherical sensors for the first iterations. Indeed, the normalized uncovered proportion reaches 70 %, with spherical sensors, after 10 iterations, while it stays under 10 % for semispherical sensors.
- The approach performs the same for both types of sensors after 30 iterations.

This can be explained by the fact that the density of sensors is the same for both types; therefore, it takes more time for the semi-spherical sensors to reduce the holes.

5.7.2 Mobility Modeling

The ALUL, denoted by $\mathcal{L}(x, \theta)$, gives an approximation of the average distance that can be made by a target, moving in 3D space, before being detected by the sensor network. The metric $\text{ALUL}_m(x, t)$ representing the ALUL in a location x at time t is given by:

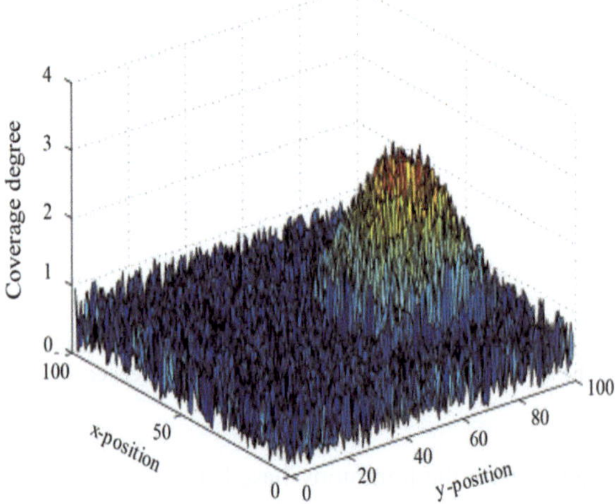

Fig. 5.2 Local node density distribution after 5 mobility iterations

$$\mathrm{ALUL}_m(x,t) = \begin{cases} 0, & \text{if} x \text{ is covered by a sensor.} \\ \frac{1}{(2\pi)^2} \int\limits_{-\frac{\pi}{2}}^{\frac{\pi}{2}} \int\limits_{0}^{2\pi} \mathcal{L}(x,\theta_1,\theta_2,t)\mathrm{d}\theta_1 \mathrm{d}\theta_2, & \text{otherwise.} \end{cases}$$

From the performance evaluation perspective, $ALUL_m(A)$ provides information on the coverage-preserving capabilities of the mobility model and the long-term behavior of the mobility model.

In order to visually illustrate the performance of coverage reduction models, we use the local node density distribution that gives the number of sensors that cover every point of the monitored region. Figure 5.2 shows that, in the simple context where one target is moving within a 100 m^2-size monitored zone, the coverage degree considerably varies according proximity to the mobile target. In fact, after 5 mobility steps, the local sensor density is less than 1 in regions that are far from the target location (which is (70, 30)) and reaches 2.7 in points that are close to this target.

More interestingly, Fig. 5.3 addresses the case where two targets are present within the region of interest. We notice that the sensors are initially uniformly distributed. The density then increases for the three following iterations in the regions where the targets are. In fact, this proves that our tracking scheme is precise enough to distinguish between the two different targets.

To confirm these results, we used the ALUL_m metric to evaluate the evolution of the uncovered area with respect to time. In fact, this allows to know whether uncovered regions are created due to the density increase in the zones that are close to the target. We compared our scheme to four known mobility models, which are the random walk model, the random way-point model, the random direction model, and the Gauss-Markov model. The results of this comparison are depicted in Fig. 5.4.

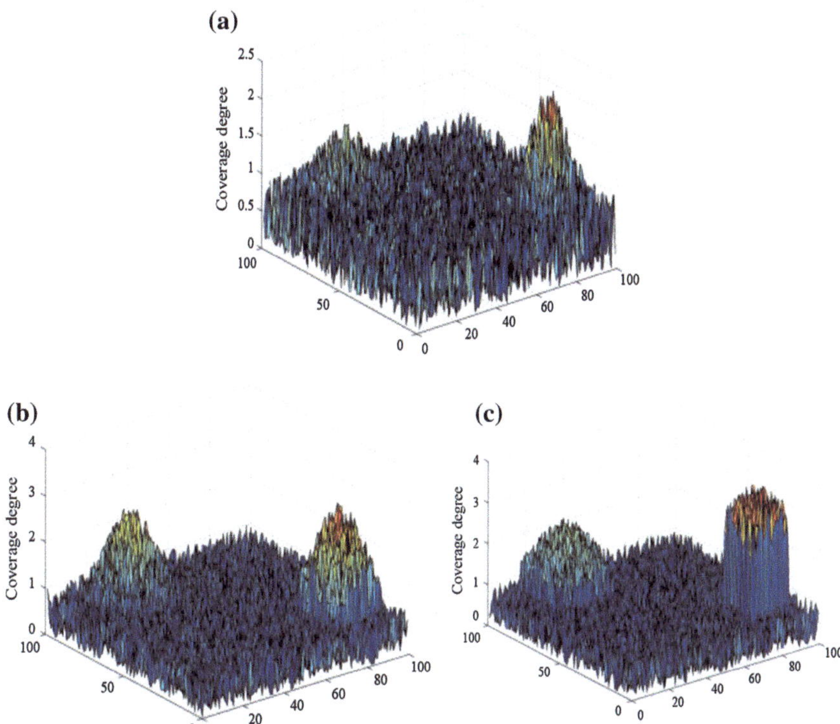

Fig. 5.3 Illustration of 3 mobility iterations for a context where two targets are considered. **a** Iteration 2, **b** Iteration 3, **c** Iteration 4

We notice that the proposed mobility models, denoted by Advanced Voronoi-based mobility model (AVBMM) and distributed Voronoi-based mobility model (DVBMM), clearly outperform the existing models. They also return a better performance than the density-preserving mobility model. This is because the latter model, despite its ability to guarantee a nearly uniform node density within the monitored area, does not take into account the presence of hostile targets in the zone of interest.

5.7.3 Complexity Evaluation

In this section, we evaluate the communication overhead resulting from the proposed retract-based coverage control approach. To this end, we only consider the complexity of the detection and localization steps in our algorithm and do not address the complexity of the repair step, since the repair step complexity is mainly dependent on the first deployment. However, one can easily deduce that if the deployment guarantees that holes size do not exceed a threshold, then the linear complexity can be verified.

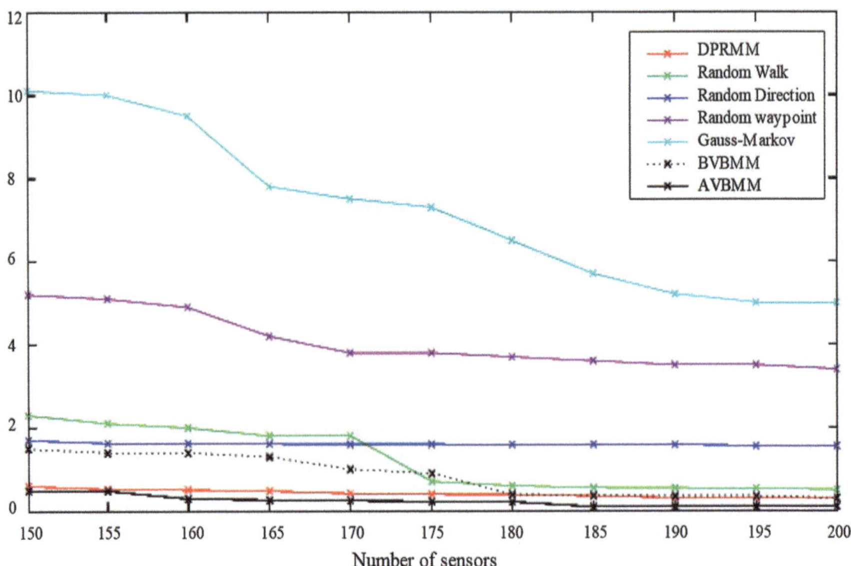

Fig. 5.4 Evolution of the ALUL$_m$ metric according to time

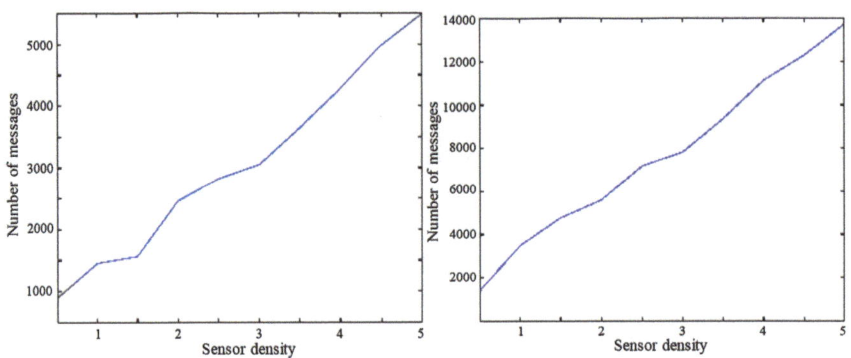

Fig. 5.5 Illustration of complexity. The *left* plot shows the first iteration, while the *right* one specifies the second iteration

We considered that the dimensions of the monitored zone are 10 m × 10 m × 3 m. We varied the number of nodes deployed within this zone, and we measured the number of messages required to set up our coverage control protocol. We first supposed that all sensor nodes have a spherical coverage of range 0.5m. Figure 5.5 depicts the number of messages for densities ranging from 0.5 sensors per m^2 to 5 sensors per m^2. The major remark is that this number is nearly linear with respect to the number of sensors per area unit.

Moreover, we considered the case where sensors have semispherical coverage (with the same range). Figure 5.5 shows that the communication overhead is also linear in this situation but with a smoother slope.

One can deduce the following statements from the aforementioned figures:

- The number of exchanged messages is independent of the density. It is close to 4 for the spherical sensors and 10 for semispherical sensors. This fact may appear strange; however, one can notice that when a deployment is performed, the detection and location will only search for holes surrounded by the Vietoris space. The latter is reduced when the density is low.
- The number of messages exchanged by the semispherical sensors for detection and localization is 2.5 times higher than the number observed for spherical. Two reasons can be mentioned for this. First, the area covered by a semispherical sensors is half the area covered by spherical sensors. Second, the guiding vectors is randomly oriented.

5.8 Summary and Outlook

This chapter developed a low-complexity approach to detect and localize sensing holes in 3D spaces. It also constructed efficient algorithms to repair holes and track (multiple targets). Our approach has built on two concepts, the Vietoris complex and the Voronoi diagram, and demonstrated that the technique called retraction by deformation achieves low-complexity algorithms for the detection of coverage holes in WSNs.

Our approach can be easily extended to more general sensors, for which the Vietoris complex and the Voronoi diagram can be defined. Such sensors can be called conical sensors or vector-guided sensors and can represent camera sensors.

References

1. C. Gui, P. Mohapatra, Power conservation and quality of surveillance in target tracking sensor networks. in *Proceedings of the 10th Annual International Conference on Mobile Computing and Networking (Mobicom 04)*, Philadelphia, PA, USA, 26 September–01 October, pp. 129–143 (2004)
2. B. Liu, D, Towsley, A study of the coverage of large-scale sensor networks. in *Proceedings of 2004 IEEE International Conference on Mobile Ad-hoc and Sensor Systems*, Fort Lauderdale, FL, USA, 25–27 October, pp. 475–483 (2004)
3. Y. Zou, K. Chakrabarty, Sensor deployment and target localization in distributed sensor networks. ACM Trans. Embed. Comput. Syst. **3**, 61–91 (2004)
4. S.P. Fekete, A. Kroller, D. Pfisterer, S. Fischer, C. Buschmann, Neighborhood-based topology recognition in sensor networks in *Proceedings of ALGOSENSORS* (Turku, Finland, July, 2004), pp. 123–136
5. S.-P. Kuo, Y.-C. Tseng, F.-J. Wu, C.-Y. Lin, A probabilistic signal-strength-based evaluation methodology for sensor network deployment. in *Proceedings of 19th International Conference*

on Advanced Information Networking and Applications, AINA 2005, vol. 1 (Taipei, Taiwan, 2005), pp. 28–30 March 2005, pp. 319–324
6. S. Ren, Q. Li, H. Wang, X. Chen, X. Zhang, A study on object tracking quality under probabilistic coverage in sensor networks. SIGMOBILE Mob. Comput. Commun. Rev. **9**, 73–76 (2005)
7. A. Muhammad, A. Jadbabaie, Dynamic coverage verification in mobile sensor networks via switched higher order Laplacians. in *Proceedings of Robotics Science and Systems* (Atlanta, GA, USA, June 2007)
8. A. Tahbaz-Salehi, A. Jadbabaie, Distributed coverage verification in sensor networks without location information. in *Proceedings of 47th IEEE Conference on Decision and Control, CDC 2008* Cancun, Mexico, 9–11 December, pp. 4170–4176 (2008)
9. P. Corke, R. Peterson, D. Rus, Localization and navigation assisted by networked cooperating sensors and robots. Int. J. Rob. Res. **24**, 771–786 (2005)
10. H. Zhang, J. Hou, Maintaining sensing coverage and connectivity in large sensor networks. Wirel. Ad Hoc Sensor Netw. **1**, 89–123 (2005)
11. C.-F. Hsin, M. Liu, Network coverage using low duty-cycled sensors: Random & coordinated sleep algorithms. in *Proceedings of The 3rd International Symposium on Information Processing in Sensor Networks, IPSN'04*, New York, pp. 433–442 (2004)
12. M. Abdelkader, M. Hamdi, N. Boudriga, Using higher-order Voronoi tessellations for WSN-based target tracking. in *Proceedings of the IASTED International Symposium, Distributed Sensor Networks, DSN'08*, Calgary, AB, Canada, 21–24 September, pp. 430–435 (2008)
13. S. Meguerdichian, F. Koushanfar, M. Potkonjak, M. Srivastava, Coverage problems in wireless ad-hoc sensor networks. in *Proceedings of INFOCOM 2001 Twentieth Annual Joint Conference of the IEEE Computer and Communications Societies*, vol. 3, Anchorage, AK, USA, 22–26 April, pp. 1380–1387 (2001)
14. X.-Y. Li, P.-J. Wan, O. Frieder, Coverage in wireless ad hoc sensor networks. Comput. IEEE Trans. **52**, 753–763 (2003)
15. C.-F. Huang, Y.-C. Tseng, The coverage problem in a wireless sensor network. in *Proceedings of The 2nd ACM International Conference on Wireless Sensor Networks and Applications, WSNA'03*, San Diego, 19 September, pp. 115–121 (2003)
16. H. Koskinen, On the coverage of a random sensor network in a bounded domain. in *Proceedings of the 16th ITC Specialist Seminar on Performance Evaluation of Wireless and Mobile Systems*, Antwerp, Belgium, 31 August-02 September, pp. 11–18 (2004)
17. R. Ghrist, A. Muhammad, Coverage and hole-detection in sensor networks via homology. in *Proceedings of the 4th International Symposium on Information Processing in Sensor Networks, IPSN'05*, Los Angeles, CA, USA, 25–27 April, pp. 254–260 (2005)
18. A. Gramain, *Topology of Surfaces; BCS Associates: Moscow* (ID, USA, 1984)
19. A. Hatcher, *Algebraic Topology* (Cambridge University Press, Cambridge, 2001)
20. V. De Silva, R. Ghrist, Coverage in sensor networks via persistent homology. Alg. Geom. Topology **7**, 339–358 (2007)
21. V. De Silva, R. Ghrist, Coordinate-free coverage in sensor networks with controlled boundaries via homology. Int. J. Rob. Res. **25**, 1205–1222 (2006)

Part III
Security Issues

Chapter 6
A Stochastic Preserving Scheme of Location Privacy

Continued advances in positioning technologies and mobile devices have increased the demand for location-based services (LBSs) in distributed sensor networks. Every LBS provider requires the sensor nodes to report their location information as the quality of service strongly depends on the information.

6.1 Introduction

One major concern in the wide deployment of LBSs is how to preserve the location privacy of the sensor nodes while providing them with a service based on their locations. LBS providers can be victims of the privacy attackers to track some sensors, or they themselves may abuse the sensors' location information for malicious purposes [1, 2]. There are two general approaches dealing with the location privacy issue in LBSs. First, we address the couple of approaches, and then, we will focus on our contribution.

6.1.1 k-Anonymity Cloaking

In this approach, instead of sending one single node's LBS request to the server, including her exact location, k-anonymity cloaking employs a trusted third party who collects k neighboring nodes' requests and sends them all together to the LBS service provider. This approach does not address the case that the node density is high; in this case, the k nodes' locations may be very close to each other, and hence, this approach will still reveal the node's location privacy to some extent. This approach was originally proposed by Gruteser and Grunwald [3]. Their work may lead to large service delay if there are not enough nodes requesting LBSs. Later on, Gedik and Liu [4] designed a joint spatial and temporal cloaking algorithm which collects k LBS requests, each from a different node in a specified cloaking area within a specified

S. S. Iyengar et al., *Mathematical Theories of Distributed Sensor Networks*,
DOI: 10.1007/978-1-4419-8420-3_6,

time period, and then sends them to the service provider. A negative point in their work is that if there are only less than k requests within the predefined time period, the nodes' requests will be blocked.

In 2009, Meyerowitz and Choudhury [5] tried to improve the service accuracy by predicting the nodes' paths and LBS queries and send the results to nodes' before they submit queries. The main drawback of their approach is the network delay occurred because of high communication overhead. For more treatment on k-anonymity cloaking, see [6–11].

6.1.2 Location Obfuscation

We can divide the solutions with this approach into two categories: solutions that preserve the location privacy of the nodes by inserting some fake LBS requests and those which deviate a node's location from the real one in her LBS request to protect her location privacy.

As examples of the solutions in the earlier category, consider the schemes proposed by Kido et al. [12], Lu et al. [13], and Duckham and Kulik [14]. In these schemes, the node generates some fake locations (dummies) using some dummy generation methods and submits the dummies and its own location to the LBS server. The server analyzes every submitted query and replies properly. The major drawback of the solutions in this category is that the server is used inefficiently and may become the system bottleneck. Additionally, nodes' location privacy is not preserved in advance.

The second category of solutions like the ones proposed by Ardagna et al. [15], Pingley et al. [16], and Damiani et al. [17] hide nodes' real locations, e.g., by submitting shifted locations. Such schemes trade service accuracy for location privacy.

Finally, in 2013, Ming Li et. al. [2] proposed a location privacy-preserving scheme called n-CD which does not use a third party and provides a trade-off between the privacy level and the system accuracy (concealing cost).

6.1.3 Our Contribution

In this chapter, we hide a node's location in an anonymity zone obtained by the Voronoi diagram of an n-gon in a 2D plane. The scheme enables the node to specify the minimum level of privacy (λ) that it desires or the maximum error tolerance (ε) that it is willing to accept when informing the services of its location. The level of privacy is defined as the probability that the node's location would not be revealed by the privacy attackers if they know the anonymity zone and the way in which the scheme is constructed. Moreover, the error tolerance is defined as the maximum Euclidean distance between any two points in the anonymity zone in which the node is located. We will theoretically find a trade-off between λ and ε in the following form: The complement of the privacy level is inversely proportional to the cube of the error tolerance, i.e.,

$$1 - \lambda = \frac{1}{\Omega(\varepsilon^3)} \tag{6.1}$$

The mentioned relation is obtained for the static node (the case addressed by all the works mentioned in the literature review).

Additionally, in this chapter, we use a random walk in the form of Brownian motion process to model the node's mobility on the plane. To the best of our knowledge, it is the first work that proposes a location privacy-preserving scheme that hides a mobile node's location and movement path in 2D plane by considering a reasonable stochastic model of his/her movement based on the Brownian motion process. We theoretically obtain the instantaneous privacy level of the mobile node. The time complexity of our approach is linear ($O(n)$ where n is the number of vertices of the initial polygon).

6.2 Problem Specification

Let c denote a mobile node in an Euclidean plane that wants to use a LBS which is offered by some service provider (say A). For example, assume that client c is interested in finding a close by five-star restaurant and server A is a search engine which helps its clients regarding their queries. The server asks any client for its current location and the places in which the client is interested (such as restaurants and cinemas). Additionally, it may ask for some criteria like how far the interested places should be.

Location Privacy Issue

An important challenge in using the mentioned LBS is how to preserve the location privacy of the clients. A good solution is that server A does not ask for the exact location of its client; instead, a subset of the Euclidean plane $S_c \subseteq \mathbb{R}^2$ containing the client's location can be sent to the server. In this way, the server can estimate the client's location and provide some service based on the estimated location. Additionally, the location privacy of the client will be preserved to some extent.

The Query Format

Regarding the aforementioned concerns, the format of the query message which client c generates and sends to the service provider A is in the following form:

$$\mathcal{Q}_c = \langle \mathrm{ID}_c, S_c, \mathcal{I}_c, \mathcal{C}_c \rangle \tag{6.2}$$

where ID_c is the client's unique ID, $S_c \subseteq \mathbb{R}^2$ denotes the client's anonymity zone in which the client is located and parameters $\mathcal{I}_c$ and $\mathcal{C}_c$, respectively, denote the list of places in which the client is interested and the criteria which help the service provider filter the query results properly.

Anonymity Zone

Considering a Cartesian coordinate system in the plane, we assume that the mobile node obtains its instantaneous coordinates $\big(x_c(t), y_c(t)\big)$ using another service (like GPS). As mentioned before, we assume that server A needs to know an estimation of the client's coordinates on the plane when the query is sent; however, because of the location privacy issues, node c does not want to reveal its exact instantaneous location represented by $\mathrm{loc}_c(t) = \big(x_c(t), y_c(t)\big)$. This urges the client to generate some "anonymity zone" $S_c \subseteq \mathbb{R}^2$ and include it in the query message (instead of sending its precise location). Assuming that the query is sent in time $t = 0$, this is the case that

$$\mathrm{loc}_c(0) \in S_c \tag{6.3}$$

Attack Model

We assume that node c occupies the area inside and on the circle of radius r and center $\mathrm{loc}_c(t)$ for any time t. As the service only knows that the client is inside the anonymity zone S_c, it guesses point G in the anonymity zone. This guess is successful if and only if[1]:

$$G \in B\big(\mathrm{loc}_c(t), r\big) \tag{6.4}$$

for every t. Assuming that node c is located in the anonymity zone S_c for every $t \leq T$, the probability that service A successfully guesses the location of node c in given time $t \leq T$, i.e., it finds node c on the plane, is equal to

$$\mathbf{Pr}\Big[G \in B\big(\mathrm{loc}_c(t), r\big)\Big|\mathrm{loc}_c(t) \in S_c\Big] \qquad \forall t$$

Definition 6.1 Considering the anonymity zone S_c, node c has the privacy level of $\lambda(t)$ at moment t, if this is the case that

[1] In this chapter, the set of points inside or on the circle $C(O, r)$ is called the ball of center O and radius r and represented by $B(O, r)$.

$$\boldsymbol{Pr}\Big[G \in B\big(\mathrm{loc}_c(t), r\big)\Big|\mathrm{loc}_c(t) \in S_c\Big] \leq 1 - \lambda(t) \tag{6.5}$$

Moreover, node c has the privacy level λ in time interval $[0, T]$, if this is the case that

$$\lambda \leq \lambda(t) \qquad \forall t \in [0, T] \tag{6.6}$$

In fact, the privacy level is a real number in $[0, 1]$ and quantifies the location privacy of the node by bounding the probability that the node's location gets revealed. The higher value of $\lambda(t)$ specifies that the location privacy is preserving with higher certainty.

Moreover, service A needs to know the node's location roughly so that the node can use the LBS. We assume that if the service obtains the node's location with higher precision, or equivalently lower error, the quality of service will increase. This precision is quantified using Definition 6.2.

Definition 6.2 The maximum error tolerance of anonymity zone S_c is represented by ε and defined in the following way:

$$\varepsilon = \sup_{Y,Z \in S_c} ||Y - Z|| \tag{6.7}$$

where $||Y - Z||$ denotes the Euclidean distance between points $Y, Z \in \mathbb{R}^2$.

If the maximum error tolerance of an anonymity zone increases, the node's location will be revealed with lower probability. In other words, there is a trade-off between the error tolerance and the privacy level of an anonymity zone. In the fifth section, we address this relation in detail.

6.3 The Proposed Scheme for Static Node

In this section, we propose a randomized algorithm to generate an initial anonymity zone S'_c for static node c which is not moving in time interval $[0, T]$, i.e.,

$$\mathrm{loc}_c(t) = O \qquad \forall t \leq T$$

where $O \in \mathbb{R}^2$ is an arbitrary point in the Euclidean plane. We will then make the main anonymity zone S_c using zone S'_c and then analyze the privacy level of the generated S_c in the next section.

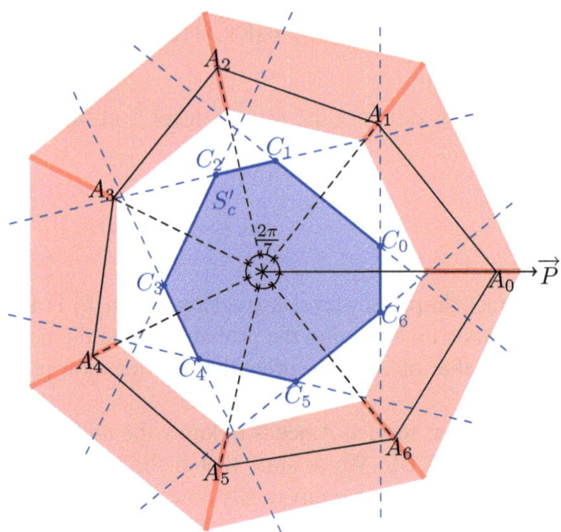

Fig. 6.1 The way that Algorithm 6.1 generates the initial anonymity zone S'_c. In this example, $n = 7$. The *red* area specifies the set of all the possible polygons created by Algorithm 6.1 in the form $\langle A_0, A_1, \ldots, A_{n-1}\rangle$. The *blue* area determines the initial anonymity zone

In Algorithm 6.1, we assume that every point is specified using a polar coordinate system of origin O and polar axis $\overrightarrow{P}$.

Algorithm 6.1: ZONEGENERATOR

Input: User's location O & odd integer $n \geq 3$ & scale factor $\mu > 0$
Output: The initial anonymity zone S'_c

for $i \leftarrow 0$ ***to*** $n - 1$ **do**
  $d \leftarrow \text{Unif}\big(\mu\cos(\frac{2\pi}{n}), \mu\big)$;
  $A_i \leftarrow$ The point of coordinates $\big(d, \frac{i\pi}{n}\big)$;
  $l_i \leftarrow$ The perpendicular bisector of $\overline{OA_i}$;
$l_n \leftarrow l_0$;
for $i \leftarrow 0$ ***to*** $n - 1$ **do**
  $C_i \leftarrow l_i \cap l_{i+1}$;
$S'_c \leftarrow$ the polygon with the sequence of vertices $\langle C_0 = C_n, C_1 = C_{n+1}, C_2, C_3, \ldots, C_{n-1}\rangle$;

As you see in Algorithm 6.1, we first draw the n-gon $\langle A_0, A_1, \ldots, A_{n-1}\rangle$ such that for every $i = 0, \ldots, n-1$, the distance between the node's location and point A_i is a uniformly random distributed variable in interval $(\mu\cos(2\pi/n), \mu)$ where μ is a positive real number called the *scale factor* and n is a positive odd integer not equal to one (see Fig. 6.1 for illustration). Then, point C_i is obtained by intersecting the perpendicular bisectors of line segments $\overline{OA_i}$[2] and $\overline{OA_{i+1}}$ (assume that $A_n = A_0$).

---

[2] The line segment between points $A \in \mathbb{R}^2$ and $B \in \mathbb{R}^2$ is represented by symbol $\overline{AB}$.

The following lemma shows that the sequence of points mentioned in line 10 of Algorithm 6.1 specifies a convex polygon of identical angles.

Lemma 6.1 *The sequence of points obtained by line 8 of Algorithm 6.1 constructs a convex n-gon containing $loc_c(t) = O$ (for every $t \le T$) such that all of its angles are identical:*

$$\measuredangle C_i C_{i+1} C_{i+2} = \pi - \frac{2\pi}{n} \qquad \forall i = 0, \ldots, n-1 \tag{6.8}$$

Proof Consider the following figure which specifies the way of finding point C_i. In this figure,

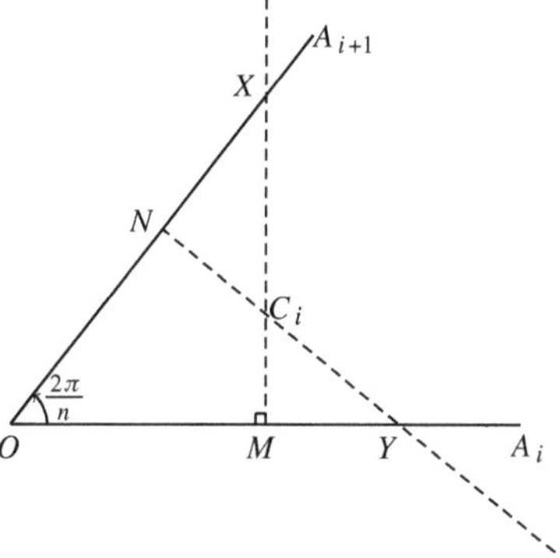

$\overline{MX} \perp \overline{OA_i}$ and $|\overline{OM}| = |\overline{MA_i}|$. Consequently, this is the case that

$$\begin{aligned}|\overline{OX}| &= \frac{|\overline{OM}|}{\cos\left(\frac{2\pi}{n}\right)} \\ &= \frac{|\overline{OA_i}|}{2\cos\left(\frac{2\pi}{n}\right)}\end{aligned}$$

As a result, regarding the second and third lines of Algorithm 6.1, we obtain the following inequalities:

$$\begin{aligned}&\mu \cos\left(\frac{2\pi}{n}\right) < |\overline{OA_i}| < \mu \\ \rightarrow &\mu \frac{\cos\left(\frac{2\pi}{n}\right)}{2\cos\left(\frac{2\pi}{n}\right)} < |\overline{OX}| < \frac{\mu}{2\cos\left(\frac{2\pi}{n}\right)}\end{aligned}$$

which implies that

$$|\overline{OX}| > \frac{\mu}{2} \tag{6.9}$$

Additionally, since $|\overline{ON}| = |\overline{OA_{i+1}}|/2$, this is the case that

$$\mu \cos\left(\frac{2\pi}{n}\right) < |\overline{OA_{i+1}}| < \mu$$
$$\rightarrow \mu \frac{\cos\left(\frac{2\pi}{n}\right)}{2} < |\overline{ON}| < \frac{\mu}{2}$$

Regarding the recent inequality and Inequality 6.9, we obtain the following relation:

$$|\overline{ON}| < |\overline{OX}| \tag{6.10}$$

Moreover, using the similar analysis, we get

$$|\overline{OM}| < |\overline{OY}| \tag{6.11}$$

Concerning Inequalities 6.10 and 6.11, we conclude that the perpendicular bisectors of line segments $\overline{OA_i}$ and $\overline{OA_{i+1}}$ meet each other at C_i. Additionally, considering quadrangle OMC_iN, this is the case that

$$\measuredangle MC_iN = 2\pi - \frac{\pi}{2} - \frac{\pi}{2} - \frac{2\pi}{n} = \pi - \frac{2\pi}{n}$$

which implies that the quadrangle OMC_iN is convex.

As the above analysis is true for every $i = 0, \ldots, n-1$, the proof is complete. □

Construction of the Anonymity Zone

Let $S'_c = \langle C_0 = C_n, C_1, \ldots, C_{n-1} \rangle$ denote the n-gon obtained by Algorithm 6.1. Additionally, assume that for every $i = 0, \ldots, n$, (x_i, y_i) represents the coordinates of point C_i in the Cartesian coordinate system of origin O and x-axis $\overrightarrow{P}$ which was mentioned before (note that $(x_0, y_0) = (x_n, y_n)$). Area $S_c \subseteq \mathbb{R}^2$ is called the anonymity zone of node c and defined in the following form:

$$\begin{aligned} S_c = \Big\{ (x, y) \in \mathbb{R}^2 \Big| \forall i = 0, \ldots, n-1 : \\ y > -h_i^{-1} x + y_i + h_i^{-1} x_i \;\; \wedge \\ y < -h_i^{-1} x + y_{i+1} + h_i^{-1} x_{i+1} \Big\} \end{aligned} \tag{6.12}$$

where

$$h_i = \frac{y_{i+1} - y_i}{x_{i+1} - x_i} \quad \forall i = 0, \ldots, n-1$$

Now, we contract some notations. Assume that lines $l_0^+, l_0^-, l_1^+, l_1^-, \ldots, l_{n-1}^+, l_{n-1}^-$ are defined in the following form:

$$l_i^+ = \{(x, y) \in \mathbb{R}^2 | y = -h_i^{-1}x + y_i + h_i^{-1}x_i\} \tag{6.13}$$

and

$$l_i^- = \{(x, y) \in \mathbb{R}^2 | y = -h_i^{-1}x + y_{i+1} + h_i^{-1}x_{i+1}\} \tag{6.14}$$

for every $i = 0, \ldots, n-1$. Note that

$$\begin{cases} l_i^+ \perp \overline{C_i C_{i+1}} \\ l_i^- \perp \overline{C_i C_{i+1}} \end{cases} \tag{6.15}$$

In addition, let X_i^+ and X_i^- represent the Euclidean distance between point O and lines l_i^+ and l_i^-, respectively (for every $i = 0, \ldots, n-1$).

Now, we prove that if we know that the static node c is located in zone S'_c, we deduce that $\mathrm{loc}_c(t) \in S_c$ for every $t \in T$, i.e., it is impossible that $\mathrm{loc}_c(t) \in S'_c - S_c$.

Lemma 6.2 *If S_c and O, respectively, denote the anonymity zone of static node c and its location in interval $[0, T]$, this is the case that*

$$O \in S_c \tag{6.16}$$

Proof Let $S'_c = \langle C_0 = C_n, C_1, \ldots, C_{n-1} \rangle$ represent the initial anonymity zone corresponding to S_c. Additionally, assume that $\mathrm{perp}(Z, \overline{C_i C_{i+1}})$ denotes the unique line passing through the arbitrary point $Z \in \mathbb{R}^2$ and perpendicular to line $\overleftrightarrow{C_i C_{i+1}}$.[3] Consider the set of points like $Z \in \mathbb{R}^2$ that hold the following condition:

$$\mathrm{perp}(Z, \overline{C_i C_{i+1}}) \cap \overleftrightarrow{C_i C_{i+1}} \in \overline{C_i C_{i+1}}$$

i.e., the set of points that the line passing through them and perpendicular to $l_i = \overleftrightarrow{C_i C_{i+1}}$ is crossing l_i at a point between C_i and C_{i+1}.

Regarding the Cartesian coordinates of every C_i, it is easy to see that

$$\begin{aligned} \mathcal{A}_i &= \{Z \in \mathbb{R}^2 | \mathrm{perp}(Z, \overline{C_i C_{i+1}}) \cap \overleftrightarrow{C_i C_{i+1}} \in \overline{C_i C_{i+1}}\} \\ &= \Big\{(x, y) \in \mathbb{R}^2 \Big| y > -h_i^{-1}x + y_i + h_i^{-1}x_i \\ &\quad \wedge y < -h_i^{-1}x + y_{i+1} + h_i^{-1}x_{i+1}\Big\} \\ &\quad \forall i = 0, \ldots, n-1 \end{aligned}$$

[3] The line passing through points $A, B \in \mathbb{R}^2$ is denoted by symbol $\overleftrightarrow{AB}$.

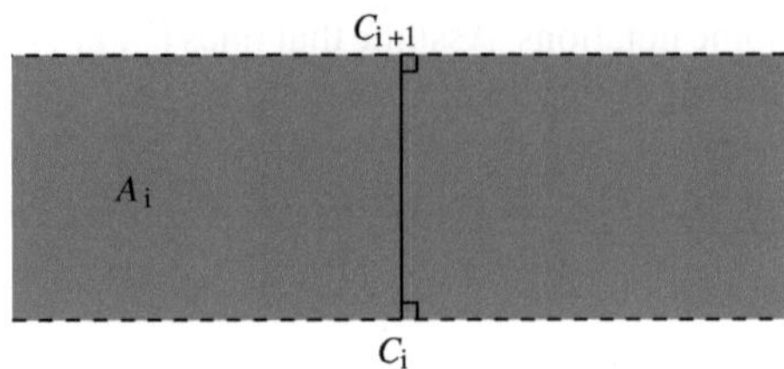

Additionally, since the midpoint of line segment $\overline{OA_{i+1}}$ is the intersection of line $\overleftrightarrow{OA_i}$ and line segment $\overline{C_iC_{i+1}}$, this is the case that

$$\text{loc}_c(t) = O \in \mathcal{A}_i \quad \forall i = 0, \ldots, n-1, t \leq T$$

or equivalently,

$$O \in \bigcap_{i=0}^{n-1} \mathcal{A}_i$$

On the other hand, concerning Eq. 6.12, area S_c can be written in the following form:

$$S' = \bigcap_{i=0}^{n-1} \mathcal{A}_i$$

which implies that

$$O \in S_c$$

□

Lemma 6.3 *Assuming that S_c' denotes the initial anonymity zone, the anonymity zone S_c obtained by Eq. 6.12 is a convex m-gon of the following form:*

$$S' = \langle D_0 = D_m, D_1, D_2, \ldots, D_m \rangle \tag{6.17}$$

such that for every $i = 0, \ldots, m-1$, $\overline{D_iD_{i+1}}$ is a line segment on line $l_i' \in L^- \cup L^+$ where

$$L^- = \{l_j^- | j = 0, \ldots, n-1\}$$

and

$$L^+ = \{l_j^+ | j = 0, \ldots, n-1\}$$

Additionally, for every $i, j = 0, \ldots, m-1$, this is the case that

$$(i \neq j) \leftrightarrow (l_i' \neq l_j') \tag{6.18}$$

Proof Proof by construction (see the following greedy algorithm). □

Here, we propose a greedy algorithm which finds the anonymity zone S_c corresponding to a given S'_c obtained by Algorithm 6.1. In this algorithm, let $\mathcal{X}$ denote the following set of random variables:

$$\mathcal{X} = \mathcal{X}^+ \cup \mathcal{X}^-$$

where

$$\mathcal{X}^+ = \{X_i^+ | i = 0, \ldots, n-1\}$$

and

$$\mathcal{X}^- = \{X_i^- | i = 0, \ldots, n-1\}$$

Also, let $\pi' : \{1, 2, \ldots, 2n\} \mapsto \mathcal{X}$ represent the increasingly sorted permutation[4] of $\mathcal{X}$ members. Additionally, assume that l_x denotes the corresponding line of random variable $x \in \mathcal{X}$ (we assume that variable X_i^+ corresponds to line l_i^+ and so on).

Phase 1: Consider lines $l_{\pi'_1}$, $l_{\pi'_2}$, and $l_{\pi'_3}$. If for every pair of lines selected out of the three, they are not parallel, we will obtain three intersecting points which make triangle $\triangle D_0D_1D_2$[5]; otherwise, we use line $l_{\pi'_4}$ to construct the convex quadrangle $D_0D_1D_2D_3$ using the intersecting points. In the earlier case, we set variable $i_0 = 4$; however, in the second one, $i_0 = 5$.
Phase 2: After constructing the triangle or the convex quadrangle, we call it as the "basic polygon" and do the following procedure for $i = i_0, i_0 + 1, \ldots, 2n$: if line $l_{\pi'_i}$ does not cross the basic polygon or cross it in a single point, start the procedure with the next i; otherwise, since line $l_{\pi'_i}$ partitions the plane into two half planes "accepted" and "rejected," mark every vertex of the basic polygon as "accepted" if it is located in the "accepted" half plane and "rejected" otherwise. Also, mark the intersections of the polygon and line $l_{\pi'_i}$ as two[6] additional accepted vertices. Reconstruct the basic polygon using the accepted vertices (including the pair of new ones).
The algorithm outputs the basic polygon after we finish running the above procedure for $i = 2n$.

See Fig. 6.2 as an illustration of how we generate the anonymity zone using the mentioned greedy algorithm.

[4] Permutation π on set M is a total, one-to-one function in the form $\pi : \{1, 2, \ldots, |M|\} \mapsto M$ such that π_i denotes the member of M corresponding to integer $i \leq |M|$.

[5] Although it is possible that the three lines meet each other in one point, the probability of such event is zero. Henceforth, we do not consider such case.

[6] As the basic polygon is convex for every $i \geq i_0$, its edges have at most two intersecting points with a straight line.

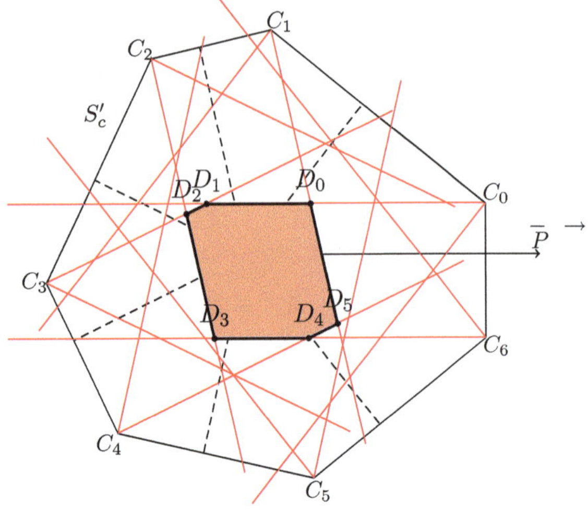

Fig. 6.2 The way of constructing the anonymity zone S_c using the initial one (S'_c) obtained by Algorithm 6.1

6.4 Location Retrieval of a Static Node

As mentioned before, service provider A does not know the exact location of the node; however, it knows that the node is located inside area S_c. Additionally, it is aware of the details of the procedure in which area S_c has been generated. This awareness makes the server guess reasonably and not choose a point where it is impossible for the node to be.

In this section, we deduce some probabilistic information regarding the location of the static node from the fact that $\text{loc}_c(t) \in S_c$ for every $t \leq T$, assuming that the node is *not* moving in time interval $[0, T]$. Since our analysis is restricted to the nonmobile node, we assume that $\text{loc}_c(t) = O$ for some point $O \in \mathbb{R}^2$.

6.4.1 Probability Distribution Over the Anonymity Zone

Let $H_i = |\overline{OM_i}|$ where $M_i \in \mathbb{R}^2$ represents the midpoint of line segment $\overline{OA_i}$ for every $i = 0, \ldots, n-1$ (note that $H_n = H_0$). Taking the second and third lines of Algorithm 6.1 into account, it is easy to see that for every $i = 0, \ldots, n-1$, H_i is a uniformly distributed random variable of the following form:

$$H_i \sim \text{Unif}\left(\frac{\mu}{2}\cos\left(\frac{2\pi}{n}\right), \frac{\mu}{2}\right) \tag{6.19}$$

Here, we present two lemmas to conclude the probability distribution of random variables X_i^+ and X_i^-.

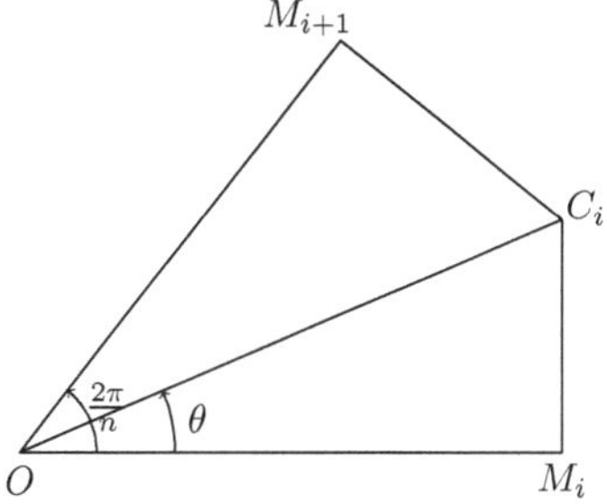

Fig. 6.3 Calculating the values of X_{i+1}^+ and X_i^- based on H_i and H_{i+1}

Lemma 6.4 *For every* $i = 0, \ldots, n-1$, *random variables* X_{i+1}^+ *and* X_i^- *can be written in the following form:*

$$X_{i+1}^+ = \frac{H_i - \cos\left(\frac{2\pi}{n}\right) H_{i+1}}{\sin\left(\frac{2\pi}{n}\right)} \quad \forall i = 0, \ldots, n-1 \tag{6.20}$$

and

$$X_i^- = \frac{H_{i+1} - \cos\left(\frac{2\pi}{n}\right) H_i}{\sin\left(\frac{2\pi}{n}\right)} \quad \forall i = 0, \ldots, n-1 \tag{6.21}$$

Note that $X_0^+ = X_n^+$ *and* $H_0 = H_n$.

Proof Consider Fig. 6.3, concerning the contracted notations, this is the case that

$$\begin{cases} |\overline{OM_i}| = H_i \\ |\overline{OM_{i+1}}| = H_{i+1} \\ |\overline{C_iM_i}| = X_i^- \\ |\overline{C_iM_{i+1}}| = X_{i+1}^+ \end{cases}$$

Since $\overline{OM_i} \perp \overline{M_iC_i}$, we obtain:

$$\begin{aligned} |\overline{OC_i}| &= \frac{|\overline{OM_i}|}{\cos(\theta)} \\ &= \frac{II_i}{\cos(\theta)} \end{aligned}$$

Additionally, as $\overline{OM_{i+1}} \perp \overline{M_{i+1}C_i}$, we get

$$\begin{aligned} |\overline{OC_i}| &= \frac{|\overline{OM_{i+1}}|}{\cos\left(\frac{2\pi}{n} - \theta\right)} \\ &= \frac{H_{i+1}}{\cos\left(\frac{2\pi}{n} - \theta\right)} \end{aligned}$$

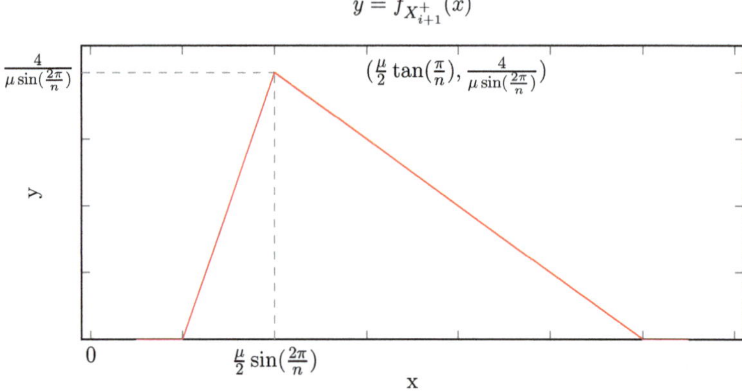

Fig. 6.4 Plot of the probability distribution function of random variable X^+_{i+1}

This leads to the following equation:

$$\frac{H_i}{\cos(\theta)} = \frac{H_{i+1}}{\cos\left(\frac{2\pi}{n} - \theta\right)}$$

By simplifying the equation, this is the case that

$$\tan(\theta) = \frac{H_{i+1} - \cos\left(\frac{2\pi}{n}\right) H_i}{\sin\left(\frac{2\pi}{n}\right) H_i}$$

Henceforth, we can find the closed form of X^-_i:

$$\begin{aligned} X^-_i &= |\overline{CM_i}| \\ &= \tan(\theta)|\overline{OM_i}| \\ &= \tan(\theta) H_i \\ &= \frac{H_{i+1} - \cos\left(\frac{2\pi}{n}\right) H_i}{\sin\left(\frac{2\pi}{n}\right)} \end{aligned}$$

Using the similar way, we can conclude the closed form of X^+_{i+1}. □

Lemma 6.5 *For every $i = 0, \ldots, n-1$, if $f_{X^+_{i+1}}$ and $f_{X^-_i}$, respectively, denote the probability distribution functions of X^+_{i+1} and X^-_i, this is the case that $f_{X^+_{i+1}}(x) = f_{X^-_i}(x)$ and Fig. 6.4 specifies the plot of $y = f_{X^+_{i+1}}(x)$:*

Proof Regarding Lemma 6.4, this is the case that

$$\begin{cases} X_{i+1}^{+} = \dfrac{H_i - \cos\left(\frac{2\pi}{n}\right) H_{i+1}}{\sin\left(\frac{2\pi}{n}\right)} \\ X_i^{-} = \dfrac{H_{i+1} - \cos\left(\frac{2\pi}{n}\right) H_i}{\sin\left(\frac{2\pi}{n}\right)} \end{cases}$$

or equivalently,

$$\begin{cases} H_i = \dfrac{X_{i+1}^{+} + \cos\left(\frac{2\pi}{n}\right) X_i^{-}}{\sin\left(\frac{2\pi}{n}\right)} \\ H_{i+1} = \dfrac{\cos\left(\frac{2\pi}{n}\right) X_{i+1}^{+} + X_i^{-}}{\sin\left(\frac{2\pi}{n}\right)} \end{cases}$$

This implies that

$$\begin{aligned} f_{X_{i+1}^{+} X_i^{-}}(x, y) &= \left|\det\left(\begin{bmatrix} \csc\left(\frac{2\pi}{n}\right) & \cot\left(\frac{2\pi}{n}\right) \\ \cot\left(\frac{2\pi}{n}\right) & \csc\left(\frac{2\pi}{n}\right) \end{bmatrix}\right)\right| \\ &\quad \times f_{H_i H_{i+1}}\left(\frac{x + \cos\left(\frac{2\pi}{n}\right) y}{\sin\left(\frac{2\pi}{n}\right)}, \frac{\cos\left(\frac{2\pi}{n}\right) x + y}{\sin\left(\frac{2\pi}{n}\right)}\right) \\ &= f_{H_i H_{i+1}}\left(\frac{x + \cos\left(\frac{2\pi}{n}\right) y}{\sin\left(\frac{2\pi}{n}\right)}, \frac{\cos\left(\frac{2\pi}{n}\right) x + y}{\sin\left(\frac{2\pi}{n}\right)}\right) \end{aligned}$$

Since H_i and H_{i+1} are independent and uniformly distributed (see Relation 6.19), this is the case that

$$f_{H_i H_{i+1}}(h_1, h_2) = \begin{cases} \dfrac{1}{\mu^2 \sin^4\left(\frac{\pi}{n}\right)} & \bar{h} \in \left(\dfrac{\mu \cos\left(\frac{2\pi}{n}\right)}{2}, \dfrac{\mu}{2}\right)^2 \\ 0 & \text{otherwise} \end{cases}$$

Consequently, the joint probability distribution function of X_{i+1}^{+} and X_i^{-} is obtained by the following equation:

$$f_{X_{i+1}^{+} X_i^{-}}(x, y) = \begin{cases} \dfrac{1}{\mu^2 \sin^4\left(\frac{\pi}{n}\right)} & (x, y) \in \mathcal{S} \\ 0 & \text{otherwise} \end{cases} \tag{6.22}$$

where $\mathcal{S} \subseteq \mathbb{R}^2$ is the subset of 2D plane in the following form:

$$\mathcal{S} = \left\{(x, y) \middle| \frac{\mu \cos\left(\frac{2\pi}{n}\right)}{2} \leq \frac{x + \cos\left(\frac{2\pi}{n}\right) y}{\sin\left(\frac{2\pi}{n}\right)} \leq \frac{\mu}{2} \wedge \frac{\mu \cos\left(\frac{2\pi}{n}\right)}{2} \leq \frac{\cos\left(\frac{2\pi}{n}\right) x + y}{\sin\left(\frac{2\pi}{n}\right)} \leq \frac{\mu}{2}\right\}$$

As we are interested in the marginal probability distribution functions, this is the case that

$$f_{X_{i+1}^{+}}(x) = \int_{-\infty}^{+\infty} f_{X_{i+1}^{+}X_{i}^{-}}(x, y)\mathrm{d}y \tag{6.23}$$

and

$$f_{X_{i}^{-}}(y) = \int_{-\infty}^{+\infty} f_{X_{i+1}^{+}X_{i}^{-}}(x, y)\mathrm{d}x \tag{6.24}$$

By simplifying Eqs. 6.23 and 6.24, we conclude the lemma. □

6.4.2 Computing the Privacy Level

As mentioned before, if service provider A guesses that node c is in point $G \in \mathbb{R}^2$, the probability that this guess is successful is equal to

$$\mathbf{Pr}\Big[G \in B\big(O, r\big)\Big| O \in S_c\Big]$$

The following lemma proposes an upper bound on this value.

Lemma 6.6 *Assuming that X_i' denotes a random variable equal to the Euclidean distance between the node's location (O) and line l_i' (which contains line segment $\overline{D_i D_{i+1}}$) for every $i = 0, \ldots, m-1$, this is the case that*

$$\begin{aligned}&\mathbf{Pr}\Big[G \in B\big(O, r\big)\Big| O \in S_c\Big]\\&\quad\le \mathbf{Pr}\Big[\bigwedge_{i=0}^{m-1} \big(X_i' \in [\delta_i - r, \delta_i + r]\big)\Big]\end{aligned}$$

where δ_i is the Euclidean distance from point G to line l_i' for every $i = 0, \ldots, m-1$.

Proof Let $\mathrm{dist}(Z, l_i')$ denote the Euclidean distance from arbitrary point $Z \in \mathbb{R}^2$ to line l_i' (for every $i = 0, \ldots, m-1$). Assuming that $G \in B(O, r)$, this is the case that

$$|\overline{OG}| \le r \tag{6.25}$$

Since $\mathrm{dist}(G, l_i') = \delta_i$ for every $i = 0, \ldots, m-1$, Inequality 6.25 and the triangular inequality lead to the following relation:

$$\delta_i - r \le \mathrm{dist}(G, l_i') \le \delta_i + r \quad \forall i = 0, \ldots, m-1$$

or equivalently,

$$\bigwedge_{i=0}^{m-1} (X'_i \in [\delta_i - r, \delta_i + r])$$

Additionally, regarding Lemma 6.2, service provider A knows that $O \in S_c$ and does not guess a point inside area $S'_c - S_c$. Henceforth, we conclude the lemma. □

Now, we focus on finding the exact value of the following probability which is a function of variables $\delta_0, \delta_1, \ldots, \delta_{m-1}$:

$$\mathbf{Pr}\Big[\bigwedge_{i=0}^{m-1} \big(X'_i \in [\delta_i - r, \delta_i + r]\big)\Big] = f(\delta_0, \delta_1, \ldots, \delta_{m-1})$$

Here is the approach of computing the value of function f in four steps:

Step 1

Rewrite the mentioned probability as a product of m terms:

$$\begin{aligned}
&\mathbf{Pr}\Big[\bigwedge_{i=0}^{m-1} \big(X'_i \in [\delta_i - r, \delta_i + r]\big)\Big] \\
&= \mathbf{Pr}\big[X'_0 \in [\delta_0 - r, \delta_0 + r]\big] \times \\
&\quad \mathbf{Pr}\big[X'_1 \in [\delta_1 - r, \delta_1 + r] \big| X'_0 \in [\delta_0 - r, \delta_0 + r]\big] \times \\
&\vdots \\
&\quad \mathbf{Pr}\big[X'_{m-1} \in [\delta_{m-1} - r, \delta_{m-1} + r] \big| \bigwedge_{i=0}^{m-2} X'_i \in [\delta_i - r, \delta_i + r]\big]
\end{aligned}$$

Step 2

Simplify the conditional probabilities obtained in the previous step using the following lemma.

Lemma 6.7 *For every $i = 0, \ldots, m-1$, random variable X'_i is independent of $Y \in \mathcal{X} - (D_{X'_i} \cup \{X'_i\})$ where set $D_{X'_i}$ is defined in the following form:*

$$D_{X'_i} = \begin{cases} \{X^+_{j+1}, X^+_{j+3}, X^-_j, X^-_{j+1}, X^-_{j+2}\} & \text{if} X'_i = X^+_{j+2} \\ \{X^-_j, X^-_{j+2}, X^+_{j+1}, X^+_{j+2}, X^+_{j+3}\} & \text{if} X'_i = X^-_{j+1} \end{cases}$$

Proof As $H_0, H_1, \ldots, H_{n-1}$ are mutually independent, we directly conclude the lemma using Eqs. 6.20 and 6.21. □

To simplify the kth conditional probability, we eliminate the conditions on those variables that do not belong to $D_{X'_k}$, i.e.,

$$\mathbf{Pr}\big[X'_k \in [\delta_k - r, \delta_k + r]\big| \bigwedge_{i=0}^{k-1} X'_i \in [\delta_i - r, \delta_i + r]\big]$$
$$= \mathbf{Pr}\big[X'_k \in [\delta_k - r, \delta_k + r]\big| \bigwedge_{\substack{i=0,\dots,k-1 \\ X'_i \in D_{X'_k}}} X'_i \in [\delta_i - r, \delta_i + r]\big]$$

Step 3
After eliminating the unnecessary conditions in Step 2, we compute the probabilities with no conditions using Lemma 6.5:

$$\mathbf{Pr}\big[X'_i \in [\delta_i - r, \delta_i + r]\big] = \int_{\delta_i - r}^{\delta_i + r} f_{X'_i}(x)\mathrm{d}x$$

Step 4
In the last step, we compute the conditional probabilities. Every conditional probability has at most $|D_{X'_i}| = 5$ conditions and is computed similar to one of these three cases (note that $\Delta_i = (\delta_i - r, \delta_i + r)$ for every $i = 0, \dots, m-1$):

(i)
$$\mathbf{Pr}\big[X^-_k \in \Delta_i | X^+_{k+1} \in \Delta_j\big] = \frac{\mathbf{Pr}\big[X^-_k \in \Delta_i \wedge X^+_{k+1} \in \Delta_j\big]}{\mathbf{Pr}\big[X^+_{k+1} \in \Delta_j\big]}$$

Probability $\mathbf{Pr}\big[X^-_k \in \Delta_i \wedge X^+_{k+1} \in \Delta_j\big]$ is computed using the joint probability distribution function obtained in Eq. 6.22. Additionally, $\mathbf{Pr}\big[X^+_{k+1} \in \Delta_j\big]$ is computed using Lemma 6.5.

(ii)
$$\mathbf{Pr}\big[X^-_k \in \Delta_i | X^+_k \in \Delta_j\big] = \frac{\mathbf{Pr}\big[X^-_k \in \Delta_i \wedge X^+_k \in \Delta_j\big]}{\mathbf{Pr}\big[X^+_k \in \Delta_j\big]}$$

Value of $\mathbf{Pr}\big[X^+_k \in \Delta_j\big]$ is computed using Lemma 6.5. In addition, regarding Eqs. 6.20 and 6.21, this is the case that

$$\mathbf{Pr}\big[X^-_k \in \Delta_i \wedge X^+_k \in \Delta_j\big] = \mathbf{Pr}\Big[\frac{H_{k+1} - \cos\left(\frac{2\pi}{n}\right) H_k}{\sin\left(\frac{2\pi}{n}\right)} \in \Delta_i \wedge \frac{H_{k-1} - \cos\left(\frac{2\pi}{n}\right) H_k}{\sin\left(\frac{2\pi}{n}\right)} \in \Delta_j\Big]$$
$$= \int_{h=\frac{\mu}{2}\cos\left(\frac{2\pi}{n}\right)}^{\frac{\mu}{2}} \int_{h'=\frac{\mu}{2}\cos\left(\frac{2\pi}{n}\right)}^{\frac{\mu}{2}} \mathbf{Pr}\Bigg[\frac{h - (\delta_i + r)\sin\left(\frac{2\pi}{n}\right)}{\cos\left(\frac{2\pi}{n}\right)} < H_k$$
$$< \frac{h - (\delta_i - r)\sin\left(\frac{2\pi}{n}\right)}{\cos\left(\frac{2\pi}{n}\right)} \wedge \frac{h' - (\delta_j + r)\sin\left(\frac{2\pi}{n}\right)}{\cos\left(\frac{2\pi}{n}\right)} < H_k$$
$$< \frac{h' - (\delta_j - r)\sin\left(\frac{2\pi}{n}\right)}{\cos\left(\frac{2\pi}{n}\right)}\Bigg] \times \left(\frac{1}{\mu \sin^2\left(\frac{\pi}{n}\right)}\right)^2 \mathrm{d}h\, \mathrm{d}h'$$

$$= \int_{h=\frac{\mu}{2}\cos\left(\frac{2\pi}{n}\right)}^{\frac{\mu}{2}} \int_{h'=\frac{\mu}{2}\cos\left(\frac{2\pi}{n}\right)}^{\frac{\mu}{2}} \int_{h''=\frac{\mu}{2}\cos\left(\frac{2\pi}{n}\right)}^{\frac{\mu}{2}} \left|\left(\max\left\{\frac{h-(\delta_i+r)\sin\left(\frac{2\pi}{n}\right)}{\cos\left(\frac{2\pi}{n}\right)}, \frac{h'-(\delta_j+r)\sin\left(\frac{2\pi}{n}\right)}{\cos\left(\frac{2\pi}{n}\right)}\right\},\right.\right.$$
$$\left.\left.\min\left\{\frac{h-(\delta_i-r)\sin\left(\frac{2\pi}{n}\right)}{\cos\left(\frac{2\pi}{n}\right)}, \frac{h'-(\delta_j-r)\sin\left(\frac{2\pi}{n}\right)}{\cos\left(\frac{2\pi}{n}\right)}\right\}\right)\right|$$
$$\times \left(\frac{1}{\mu\sin^2\left(\frac{\pi}{n}\right)}\right)^3 \mathrm{d}h\,\mathrm{d}h'\,\mathrm{d}h''$$

(iii)

$$\mathbf{Pr}\left[X_k^- \in \Delta_i \middle| (X_k^+, X_{k+2}^+) \in \Delta_p \times \Delta_q\right] = \frac{\mathbf{Pr}\left[(X_k^-, X_k^+, X_{k+2}^+) \in \Delta_i \times \Delta_p \times \Delta_q\right]}{\mathbf{Pr}\left[(X_k^+, X_{k+2}^+) \in \Delta_p \times \Delta_q\right]}$$
$$= \frac{\mathbf{Pr}\left[X_k^- \in \Delta_i\right]\mathbf{Pr}\left[X_k^+ \in \Delta_p \middle| X_k^- \in \Delta_i\right]}{\mathbf{Pr}\left[(X_k^+) \in \Delta_p\right]} \times \frac{\mathbf{Pr}\left[X_{k+2}^+ \in \Delta_q \middle| (X_k^-, X_k^+) \in \Delta_i \times \Delta_p\right]}{\mathbf{Pr}\left[X_{k+2}^+ \in \Delta_q\right]}$$

Additionally, since random variables X_{k+2}^+ and X_k^+ are independent, this is the case that

$$\mathbf{Pr}\left[X_{k+2}^+ \in \Delta_q \middle| (X_k^-, X_k^+) \in \Delta_i \times \Delta_p\right] = \mathbf{Pr}\left[X_{k+2}^+ \in \Delta_q \middle| X_k^- \in \Delta_i\right]$$

The result of above calculations can be simplified using the methods described in the first and second cases.

After finding function $f(\delta_0, \delta_1, \ldots, \delta_{m-1})$, the privacy level of the static node's location is obtained in the following way:

$$f(\delta_0, \delta_1, \ldots, \delta_{m-1}) = \mathbf{Pr}\left[\bigwedge_{i=0}^{m-1} \left(X_i' \in [\delta_i - r, \delta_i + r]\right)\right]$$
$$\xrightarrow{\text{Lemma 6.6}}$$
$$f(\delta_0, \delta_1, \ldots, \delta_{m-1}) \geq \mathbf{Pr}\left[G \in B(O, r) \middle| O \in S_c\right]$$
$$\xrightarrow{\text{Definition 6.1}}$$
$$\lambda \geq 1 - f(\delta_0, \delta_1, \ldots, \delta_{m-1})$$

Regarding the described approach which has four steps, the time complexity of calculating the value of function $f(\delta_0, \delta_1, \ldots, \delta_{m-1})$ is $O(m) = O(n)$.

6.5 Privacy Level Versus Error Tolerance

As mentioned before, there is a trade-off between the values of λ and ε. In this section, we first bound the values of λ and ε for a static node. Then, we obtain a relation between their values, which specifies the trade-off between them.

6.5.1 Upper Bound on the Maximum Error Tolerance

Now, we present a lemma to bind the value of maximum error tolerance for the anonymity zone S_c which was described previously:

Lemma 6.8 *Assuming that the node c is located in point $O \in \mathbb{R}^2$ in time interval $[0, T]$, this is the case that*

$$S_c \subseteq B\left(O, \mu \sin\left(\frac{\pi}{n}\right)\right) \tag{6.26}$$

Proof Regarding the definition of the anonymity zone in Eq. 6.12, we claim the following relation:

$$S_c \subseteq \mathcal{M} = \left\{(x, y) \in \mathbb{R}^2 \middle| \forall i = 0, \ldots, n-1 : y > w_i x + z_i \wedge y < w_i x + z_i'\right\} \tag{6.27}$$

where line $L_i^+ : y = w_i x + z_i$ ($L_i^- : y = w_i x + z_i'$) is parallel to line l_i^+ (l_i^-) and the Euclidean distance between the node's location and L_i^+ (L_i^-) is $\frac{\mu}{2} \sin\left(\frac{2\pi}{n}\right)$ ($L_i^+ \neq L_i^-$).

In order to prove the claim (Relation 6.27), note that since $X_i^+ \le \frac{\mu}{2} \sin\left(\frac{2\pi}{n}\right)$ and $X_i^- \le \frac{\mu}{2} \sin\left(\frac{2\pi}{n}\right)$, the deterministic lines L_i^+ and L_i^- are, respectively, the extreme cases of the stochastic lines l_i^+ and l_i^-.

Because of the symmetry, polygon $\mathcal{M}$ is a regular $2n$-gon which has an inscribed circle of radius $\frac{\mu}{2} \sin\left(\frac{2\pi}{n}\right)$. Now, we use Fig. 6.5 to compute the radius of the polygon circumcircle:

In Fig. 6.5, we obtain the following equation:

$$|\overline{OM}| = \frac{|\overline{OH}|}{\cos\left(\frac{\pi}{n}\right)}$$

Since the inscribed circle is $C\left(O, \frac{\mu}{2} \sin\left(\frac{2\pi}{n}\right)\right)$, this is the case that

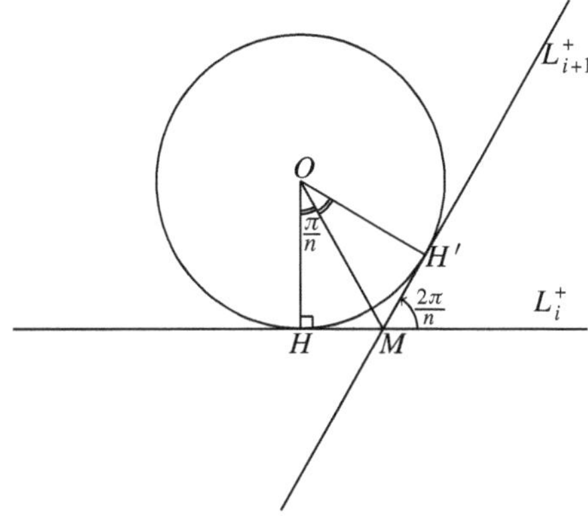

Fig. 6.5 Illustration of lines L^+_i and L^+_{i+1} which are used to prove Lemma 6.8

$$|\overline{OH}| = \frac{\mu}{2}\sin\left(\frac{2\pi}{n}\right)$$
$$\rightarrow|\overline{OM}| = \mu\sin\left(\frac{\pi}{n}\right)$$

Because of the symmetry, the value of $|\overline{OM}|$ is equal to the circumradius[7] of polygon $\mathcal{M}$. Consequently, we conclude that

$$S_c \subseteq \mathcal{M} \subseteq B\left(O, \mu\sin\left(\frac{\pi}{n}\right)\right)$$

□

Lemma 6.8 implies that

$$\sup_{Y,Z\in S_c} ||Y - Z|| \leq 2\mu\sin\left(\frac{\pi}{n}\right)$$
$$\text{Definition 6.2} \quad \varepsilon \leq 2\mu\sin\left(\frac{\pi}{n}\right)$$

6.5.2 *Lower Bound for the Privacy Level Value*

Consider Eq. 6.20 again. For every $i = 0, \ldots, n-1$, we can write H_i as a linear function of the following form:

$$H_i = \sin\left(\frac{2\pi}{n}\right)\frac{\sum_{j=i}^{n+i-1}(k_n)^{j-i}X^+_{j+1}}{1-(k_n)^n} \quad \forall i = 0, \ldots, n-1 \tag{6.28}$$

[7] Radius of the circumscribed circle.

such that

$$X_{j+n}^{+} = X_j^{+} \quad \forall j = 0, \ldots, n-1 \tag{6.29}$$

and

$$k_n = \cos\left(\frac{2\pi}{n}\right) \sin\left(\frac{2\pi}{n}\right)$$

Since for every $i = 0, \ldots, n-1$, H_i is equal to $|\overline{OA_i}|/2$, and the length of line segment $\overline{OA_i}$ is obtained by the second line of Algorithm 6.1, variables $H_0, H_1, \ldots, H_{n-1}$ are mutually independent. Consequently, the joint probability distribution function of these random variables is obtained by the following equation:

$$f_{H_0,H_1,\ldots,H_{n-1}}(h_0, h_1, \ldots, h_{n-1}) = \prod_{i=0}^{n-1} f_{H_i}(h_i) \tag{6.30}$$

Moreover, regarding Relation 6.19, every H_i is uniformly distributed. Assuming that f_u represents the probability distribution function of the uniform random variable mentioned in Relation 6.19, Eq. 6.30 will be simplified in the following form:

$$f_{H_0,H_1,\ldots,H_{n-1}}(h_0, h_1, \ldots, h_{n-1}) = \prod_{i=0}^{n-1} f_u(h_i) \tag{6.31}$$

As we have already computed every H_i in the form of a linear function of $X_0^+, X_1^+, \ldots, X_{n-1}^+$ (see Eq. 6.28), we obtain the following joint probability distribution function:

$$\begin{aligned} & f_{X_0^+,X_1^+,\ldots,X_{n-1}^+}(x_0, x_1, \ldots, x_{n-1}) \\ & \quad = |J_{\mathcal{F}}| \cdot f_{H_0,H_1,\ldots,H_{n-1}}(y_0, y_1, \ldots, y_{n-1}) \end{aligned} \tag{6.32}$$

such that $|J_{\mathcal{F}}|$ is the absolute value of the determinant of the following Jacobin matrix:

$$J_{\mathcal{F}} = \begin{bmatrix} \frac{\partial \mathcal{F}_0}{\partial x_0} & \frac{\partial \mathcal{F}_0}{\partial x_1} & \cdots & \frac{\partial \mathcal{F}_0}{\partial x_{n-1}} \\ \frac{\partial \mathcal{F}_1}{\partial x_0} & \frac{\partial \mathcal{F}_1}{\partial x_1} & \cdots & \frac{\partial \mathcal{F}_1}{\partial x_{n-1}} \\ \vdots & \vdots & \ddots & \vdots \\ \frac{\partial \mathcal{F}_{n-1}}{\partial x_0} & \frac{\partial \mathcal{F}_{n-1}}{\partial x_1} & \cdots & \frac{\partial \mathcal{F}_{n-1}}{\partial x_{n-1}} \end{bmatrix} \tag{6.33}$$

where

$$\begin{aligned} y_i &= \mathcal{F}_i(x_0 = x_n, x_1 = x_{n+1}, \ldots, x_{n-1} = x_{2n-1}) \\ &= \frac{\sin\left(\frac{2\pi}{n}\right)}{1-(k_n)^n} \sum_{j=i}^{n+i-1} (k_n)^{j-i} x_{j+1} \quad \forall i = 0, \ldots, n-1 \end{aligned} \tag{6.34}$$

and

$$x_{j+n} = x_j \quad \forall j = 0, \ldots, n-1 \tag{6.35}$$

By replacing the mutual probability distribution function of variables $H_0, H_1, \ldots, H_{n-1}$ in Eq. 6.32 with its equivalent expression, we obtain the following relation:

$$\begin{aligned} & f_{X_0^+, X_1^+, \ldots, X_{n-1}^+}(x_0, x_1, \ldots, x_{n-1}) \\ &= |J_{\mathcal{F}}| \times \prod_{i=0}^{n-1} f_u \left(\frac{\sin\left(\frac{2\pi}{n}\right)}{1-(k_n)^n} \sum_{j=i}^{n+i-1} (k_n)^{j-i} x_{j+1} \right) \end{aligned} \tag{6.36}$$

Moreover, regarding Eq. 6.34, matrix $J_{\mathcal{F}}$ is simplified in the following form:

$$J_{\mathcal{F}} = [J_{ij}]_{n \times n} \tag{6.37}$$

where

$$J_{ij} = \begin{cases} \dfrac{\sin\left(\frac{2\pi}{n}\right)(k_n)^{j-i-1}}{1-(k_n)^n} & i < j \\ \dfrac{\sin\left(\frac{2\pi}{n}\right)(k_n)^{n+j-i-1}}{1-(k_n)^n} & i \geq j \end{cases} \tag{6.38}$$

where k_n is defined as before. Regarding Eqs. 6.37 and 6.38, it is easy to see that

$$\begin{aligned} |J_{\mathcal{F}}| &= \left(\frac{\sin\left(\frac{2\pi}{n}\right)}{1-(k_n)^n} \right)^n \times \left| \det \begin{bmatrix} (k_n)^{n-1} & 1 & \cdots & (k_n)^{n-2} \\ (k_n)^{n-2} & (k_n)^{n-1} & \cdots & (k_n)^{n-3} \\ \vdots & \vdots & \ddots & \vdots \\ 1 & k_n & \cdots & (k_n)^{n-1} \end{bmatrix} \right| \\ &= \left(\frac{\sin\left(\frac{2\pi}{n}\right)}{1-(k_n)^n} \right)^n \times \left(1-(k_n)^n\right)^{n-1} \\ &= \frac{\sin^n\left(\frac{2\pi}{n}\right)}{1-(k_n)^n} \\ &= \frac{\sin^n\left(\frac{2\pi}{n}\right)}{1-\sin^n\left(\frac{2\pi}{n}\right)\cos^n\left(\frac{2\pi}{n}\right)} \end{aligned}$$

As you see, the value of $|J_{\mathcal{F}}|$ in Eq. 6.36 does not depend on any of the variables $x_0, x_1, \ldots, x_{n-1}$, i.e.,

$$|J_{\mathcal{F}}| = \mathcal{J}_n = \frac{\sin^n\left(\frac{2\pi}{n}\right)}{1-\sin^n\left(\frac{2\pi}{n}\right)\cos^n\left(\frac{2\pi}{n}\right)} \tag{6.39}$$

Additionally, concerning Eqs. 6.21 and 6.28, every X_i^- can be written as the following linear function:

$$\begin{aligned} X_i^- &= \mathcal{G}_i(X_0^+, X_1^+, \dots, X_{n-1}^+) \\ &= \frac{\left((k_n)^{n-1} - \cos\left(\frac{2\pi}{n}\right)\right)X_{j+1}^+}{1-(k_n)^n} \\ &+ \frac{\sum_{q=j+1}^{n+j-1} (k_n)^{q-j-1}\left(1 - k_n\cos\left(\frac{2\pi}{n}\right)\right)X_{q+1}^+}{1-(k_n)^n} \end{aligned} \tag{6.40}$$

for every $i = 0, \dots, n-1$.

Now, we present a theorem which proposes a lower bound for the privacy level value of a static node.

Theorem 6.1 *If λ denotes the privacy level of the static node in time interval $[0, T]$, this is the case that*

$$\lambda \geq \begin{cases} 1 - a_n\left(\frac{r}{\mu}\right)^n & n \leq m \\ 1 - b_n\left(\frac{r}{\mu}\right)^m & \text{otherwise} \end{cases} \tag{6.41}$$

for some real sequences a_n and b_n.

Proof As for every $i = 0, \dots, n-1$, X_i' is the distance of the node's location from line l_i' and line l_i' belongs to set $L^+ \cup L^-$, this is the case that

$$\{X_i' | i = 0, \dots, m\} \subseteq \left(\{X_j^+ | j = 0, \dots, n-1\} \cup \{X_j^- | j = 0, \dots, n-1\}\right) \tag{6.42}$$

In other words, for every $i = 0, \dots, m-1$, there exists some $j = 0, \dots, n-1$ such that

$$X_i' = X_j^+ \tag{6.43}$$

or,

$$X_i' = X_j^- = \mathcal{G}_j(X_0^+, X_1^+, \dots, X_{n-1}^+) \tag{6.44}$$

Henceforth, we can write every X_i' as a linear function of variables $X_0^+, X_1^+, \dots, X_{n-1}^+$. This implies that

$$\bigwedge_{i=0}^{m-1} \left(X_i' \in [\delta_i - r, \delta_i + r]\right) \leftrightarrow (X_0^+, X_1^+, \dots, X_{n-1}^+) \in \mathcal{A} \tag{6.45}$$

where $\mathcal{A} \in \mathbb{R}^n$ is a n-dimensional polyhedron obtained using Eq. 6.43, 6.44, and 6.45. Consequently, this is the case that

$$\mathbf{Pr}\Big[\bigwedge_{i=0}^{m-1}\big(X'_i \in [\delta_i - r, \delta_i + r]\big)\Big] = \mathbf{Pr}\Big[(X_0^+, X_1^+, \ldots, X_{n-1}^+) \in \mathcal{A}\Big]$$

Regarding Eq. 6.36 which specifies the joint probability distribution function of variables $X_0^+, X_1^+, \ldots, X_{n-1}^+$, we obtain the following equation:

$$\begin{aligned}
&\mathbf{Pr}\Big[(X_0^+, X_1^+, \ldots, X_{n-1}^+) \in \mathcal{A}\Big] \\
&= \underset{(x_0,x_1,\ldots,x_{n-1})\in\mathcal{A}}{\int\cdots\int} f_{X_0^+,X_1^+,\ldots,X_{n-1}^+}(x_0, x_1, \ldots, x_{n-1})\mathrm{d}x_0\mathrm{d}x_1, \ldots, \mathrm{d}x_{n-1} \\
&= \underset{(x_0,x_1,\ldots,x_{n-1})\in\mathcal{A}}{\int\cdots\int} \mathcal{J}_n \cdot \prod_{i=0}^{n-1} f_u(y_i)\mathrm{d}x_0\mathrm{d}x_1, \ldots, \mathrm{d}x_{n-1}
\end{aligned}$$

Consequently,

$$\mathbf{Pr}\Big[\bigwedge_{i=0}^{m-1}\big(X'_i \in [\delta_i - r, \delta_i + r]\big)\Big] = \underset{(x_0,x_1,\ldots,x_{n-1})\in\mathcal{A}}{\int\cdots\int} \mathcal{J}_n \cdot \prod_{i=0}^{n-1} f_u(y_i)\mathrm{d}x_0\mathrm{d}x_1, \ldots, \mathrm{d}x_{n-1}$$

Regarding Lemma 6.6 and the recent equations, we obtain the following inequality:

$$\begin{aligned}
\mathbf{Pr}\Big[G \in B(O, r)\Big| O \in S_c\Big] &\le \mathbf{Pr}\Big[\bigwedge_{i=0}^{m-1}\big(X'_i \in [\delta_i - r, \delta_i + r]\big)\Big] \\
&\le \underset{(x_0,x_1,\ldots,x_{n-1})\in\mathcal{A}}{\int\cdots\int} \mathcal{J}_n \cdot \prod_{i=0}^{n-1} f_u(y_i)\mathrm{d}x_0\mathrm{d}x_1, \ldots, \mathrm{d}x_{n-1} \\
&\le \mathcal{J}_n \cdot \underset{(x_0,x_1,\ldots,x_{n-1})\in\mathcal{A}}{\int\cdots\int} \prod_{i=0}^{n-1} f_u(y_i)\mathrm{d}x_0\mathrm{d}x_1, \ldots, \mathrm{d}x_{n-1}
\end{aligned}$$

The recent inequality proposes the following privacy level for the static node:

$$\lambda \ge 1 - \mathcal{J}_n \cdot \underset{(x_0,x_1,\ldots,x_{n-1})\in\mathcal{A}}{\int\cdots\int} \prod_{i=0}^{n-1} f_u(y_i)\mathrm{d}x_0\mathrm{d}x_1, \ldots, \mathrm{d}x_{n-1} \tag{6.46}$$

where

$$f_u(x) = \begin{cases} \dfrac{2}{\mu\big(1 - \cos\big(\frac{2\pi}{n}\big)\big)} & x \in \big(\frac{\mu}{2}\cos\big(\frac{2\pi}{n}\big), \frac{\mu}{2}\big) \\ 0 & \text{otherwise} \end{cases} \tag{6.47}$$

By replacing the value of $f_u(x)$ in Inequality 6.46, we will get the following simpler form:

$$\lambda \geq 1 - \mathcal{J}_n \left(\frac{1}{\mu \sin^2\left(\frac{\pi}{n}\right)} \right)^n |\mathcal{A} \cap \mathcal{B}| \tag{6.48}$$

where $\mathcal{B} \in \mathbb{R}^n$ is defined in the following form:

$$\mathcal{B} = \left\{ \bar{x} \middle| \forall i = 0, \ldots, n-1 \colon \frac{\mu}{2} \cos\left(\frac{2\pi}{n}\right) < \mathcal{F}_i(\bar{x}) < \frac{\mu}{2} \right\}$$

and $\mathcal{A}$ is obtained by m pairs of inequalities in either of these two forms:

$$\forall i = 0, \ldots, m-1 \colon \begin{cases} \delta_i - r \leq x_j \leq \delta_i + r & \text{if } X'_i = X^+_j \\ \delta_i - r \leq \mathcal{G}_j(\bar{x}) \leq \delta_i + r & \text{if } X'_i = X^-_j \end{cases}$$

where each pair specifies the area between two parallel n-dimensional planes.

Now, assume that $n \leq m$. Polyhedron $\mathcal{A}$ is the intersection of m areas such that each one is the area between two parallel n-dimensional planes (if $X'_i = X^+_j$, the parallel planes are $P^+_i : x_j \geq \delta_i - r$ and $P^-_i : x_j \leq \delta_i + r$; otherwise, $P^+_i : \mathcal{G}_j(\bar{x}) \geq \delta_i - r$ and $P^-_i : \mathcal{G}_j(\bar{x}) \leq \delta_i + r$). Since $\mathcal{G}_i$ is a linear function, the distance between the parallel planes is a factor of r. As the result,

$$|\mathcal{A}| \leq a'_n r^n$$

which implies that

$$\begin{aligned} 1 - \lambda &\leq \mathcal{J}_n \left(\frac{1}{\mu \sin^2\left(\frac{\pi}{n}\right)} \right)^n |\mathcal{A} \cap \mathcal{B}| \\ &\leq \mathcal{J}_n \left(\frac{1}{\mu \sin^2\left(\frac{\pi}{n}\right)} \right)^n |\mathcal{A}| \\ &\leq \mathcal{J}_n \left(\frac{1}{\mu \sin^2\left(\frac{\pi}{n}\right)} \right)^n a'_n r^n \\ &\leq a_n \left(\frac{r}{\mu} \right)^n \end{aligned}$$

Now, assume the case that $n > m$. Let $\mathcal{X}'$ denote the following set of random variables:

$$\mathcal{X}' = \{X \in \mathcal{X}^+ | \forall i = 0, \ldots, m-1 : X'_i \neq X\} \tag{6.49}$$

we define n-dimensional polyhedron $\mathcal{A}' \subseteq \mathbb{R}^n$ in this form: $\bar{x} \in \mathcal{A}'$ if and only if $\bar{x} \in \mathcal{A}$ and

$$0 \leq x_i \leq \mu \sin^2\left(\frac{\pi}{n}\right) \quad \text{if } X_i^+ \in \mathcal{X}' \tag{6.50}$$

Using the similar deduction that we did for $n \leq m$, we obtain that

$$|\mathcal{A}'| \leq b_n' r^m \mu^{n-m}$$

which implies that

$$\begin{aligned}
1 - \lambda &\leq \mathcal{J}_n \left(\frac{1}{\mu \sin^2\left(\frac{\pi}{n}\right)}\right)^n |\mathcal{A} \cap \mathcal{B}| \\
&\leq \mathcal{J}_n \left(\frac{1}{\mu \sin^2\left(\frac{\pi}{n}\right)}\right)^n |\mathcal{A}'| \\
&\leq \mathcal{J}_n \left(\frac{1}{\mu \sin^2\left(\frac{\pi}{n}\right)}\right)^n b_n' r^m \mu^{n-m} \\
&\leq b_n \left(\frac{r}{\mu}\right)^m
\end{aligned}$$

□

Regarding Theorem 6.1, if we assume that $r \leq \mu$, since $m \geq 3$ and $n \geq 3$, this is the case that

$$1 - \lambda \leq e_n \left(\frac{r}{\mu}\right)^3 \tag{6.51}$$

for some real positive sequence e_n.

6.5.3 Trade-Off Between the Privacy Level and the Error Tolerance

In this section, we obtained the following inequality for the maximum error tolerance of anonymity zone S_c:

$$\varepsilon \leq 2\mu \sin\left(\frac{\pi}{n}\right)$$

which implies that

$$\mu \geq \frac{\varepsilon}{2 \sin\left(\frac{\pi}{n}\right)}$$

Additionally, regarding the computed lower bound for the value of the privacy level in Inequality 6.51, we obtain that

$$1-\lambda \leq b_n \left(\frac{r}{\mu}\right)^3$$
$$\leq b_n \left(\frac{r}{\frac{\varepsilon}{2\sin(\frac{\pi}{n})}}\right)^3$$

which implies that

$$1-\lambda \leq \eta_n \left(\frac{r}{\varepsilon}\right)^3 \tag{6.52}$$

Inequality 6.52 specifies the trade-off between the maximum error tolerance and the privacy level of the anonymity zone.

As a numerical example, if we need to increase the low threshold of the privacy level from 99 to 99.9 %, we have to increase the maximum error tolerance by factor of $\sqrt[3]{10} \simeq 2.15$. *In order to do this, we only need to multiply the value of scale factor* μ *by* 2.15.

6.6 Location Privacy Preserving of a Mobile Node

In this section, an extension of our scheme to the mobile node will be proposed. We restrict our consideration to the case that the node has subsequent random close by destinations. More precisely, the node has a sequence of independently chosen random destinations in the form $\mathcal{D}_0 = \text{loc}_c(0), \mathcal{D}_1, \mathcal{D}_2, \mathcal{D}_3, \ldots$ such that for every $i \geq 0$, the node moves from point $\mathcal{D}_i \in \mathbb{R}^2$ to $\mathcal{D}_{i+1} \in \mathbb{R}^2$ in time interval $[t_i, t_{i+1})$ such that $t_0 = 0$,

$$0 < t_{i+1} - t_i \leq \epsilon \quad \forall i = 0, 1, \ldots$$

for some small real number $\epsilon > 0$, and assuming that the maximum speed of the node is represented by M_s, this is the case that

$$\epsilon M_s \ll \mu$$

or equivalently, the distance between two consequent destinations $||\mathcal{D}_{i+1} - \mathcal{D}_i||$ is negligible (compared to the scale factor).

6.6.1 A Stochastic Model of the Node Movement

We define random processes x_t and y_t in the following form:

$$\begin{cases} x_t = x_c(t) - x_c(0) \\ y_t = y_c(t) - y_c(0) \end{cases} \quad \forall t \geq 0 \tag{6.53}$$

where $(x_c(t), y_c(t))$ denotes the Cartesian coordinates of the node's location in time t ($\text{loc}_c(t)$).

Considering the aforementioned assumptions regarding the node movement on the plane, we estimate x_t and y_t using random processes $\hat{x}_t$ and $\hat{y}_t$ which have the following properties:

1. Maps $g: t \mapsto \hat{x}_t(\omega)$ and $g': t \mapsto \hat{y}_t(\omega)$ are continuous for every ω and $t > 0$.
2. For every $k \in \mathbb{N}$, assuming that $0 \leq t_1 \leq t_2 \leq \cdots \leq t_k$, the random variables belonging to the following set are mutually independent:

$$\{(\hat{x}_{t_{i+1}} - \hat{x}_{t_i})| \;\; i = 1, \ldots, k-1\}$$

The same proposition is true for the following set:

$$\{(\hat{y}_{t_{i+1}} - \hat{y}_{t_i})| i = 1 \ldots k-1\}$$

3. Every increment of processes $\hat{x}_t$ and $\hat{y}_t$ is stationary, i.e., the probability distributions of $\hat{x}_t - \hat{x}_s$ and $\hat{y}_t - \hat{y}_s$ only depend on $t - s$ for every $t, s > 0$.

Note that since $\text{loc}_c(t) = (x_t + x_c(0), y_t + y_c(0))$, the first mentioned property is also true for x_t and y_t:

$$\begin{cases} x_t = \lim\limits_{w \to t^+} x_w = \lim\limits_{w \to t^-} x_w \\ y_t = \lim\limits_{w \to t^+} y_w = \lim\limits_{w \to t^-} y_w \end{cases}$$

However, the other two properties are reasonably estimated regarding processes x_t and y_t.

Random processes $\hat{x}$ and $\hat{y}$ are known as Brownian motion processes, and this is the case that

$$\begin{cases} \hat{x}_t \sim \mathcal{N}(0, \sigma^2 t) \\ \hat{y}_t \sim \mathcal{N}(0, \sigma^2 t) \end{cases} \tag{6.54}$$

Regarding Eq. 6.53, we find an estimation stochastic process for $x_c(t)$ and $y_c(t)$:

$$\begin{cases} x_c(t) \sim \mathcal{N}(x_c(0), \sigma^2 t) \\ y_c(t) \sim \mathcal{N}(y_c(0), \sigma^2 t) \end{cases} \tag{6.55}$$

6.6.2 Proposed Scheme for a Mobile Node

Now, we extend our scheme to the case that node c is assumed to be moving in the way mentioned previously.

Algorithm 6.2 proposes an appropriate procedure which generates an anonymity zone for mobile node c at time $t = 0$ and keeps it updated as the node is moving for every $t \geq 0$. In this procedure, we assume that function AZGENERATOR(O, n, μ, γ_n) works in the following way: Consider Algorithm 6.1 which generates the initial anonymity zone for a static node. If we change the second line of this algorithm to the form of Expression 6.56, we obtain another function called MOBILEZONEGENERATOR(O, n, μ, γ_n) where γ_n is a positive real number less than $2\sin^2\left(\frac{\pi}{n}\right)$. Function AZGENERATOR returns the zone generated by applying the greedy algorithm mentioned in Section three on the output zone of function MOBILEZONEGENERATOR.

$$d \leftarrow \mathrm{Unif}\left(\mu\left(\cos\left(\frac{2\pi}{n}\right) + \gamma_n\right), \mu\right) \tag{6.56}$$

Algorithm 6.2: MOBILEZONEUPDATER

Input: privacy level λ & odd integer $n \geq 3$ & scale factor $\mu > 0$ & $\gamma_n < 2\sin^2\left(\frac{\pi}{n}\right)$

$S_c \leftarrow$ AZGENERATOR$(\mathrm{loc}_c(0), n, \mu, \gamma_n)$;
while ***true*** **do**
  $t \leftarrow$ Now();
  //Function Now() returns the current time $t \geq 0$.
  if $loc_c(t) \in S_c$ **then**
    continue ;
  $S_c \leftarrow$ AZGENERATOR$(\mathrm{loc}_c(t), n, \mu, \gamma_n)$;

6.6.3 Computing the Instantaneous Privacy Level

Now, we find a lower bound for the node's privacy level at any given time $t > 0$. Without loss of generality, we assume that the anonymity zone S_c has been generated at time $t = 0$ and kept unchanged in time interval $[0, T]$.

$$\begin{aligned}\mathbf{Pr}\Big[G \in B\big(\mathrm{loc}_c(t), r\big)\Big|\mathrm{loc}_c(t) \in S_c\Big] = \iint_{\substack{(x_0, y_0)\\ \in S_c}} \mathbf{Pr}\Big[G \in B\big(\mathrm{loc}_c(t), r\big)\Big|\mathrm{loc}_c(0) \\ = (x_0, y_0) \wedge \mathrm{loc}_c(t) \in S_c\Big] \\ \times \mathbf{Pr}\big[\mathrm{loc}_c(0) = (x_0, y_0)\big]\mathrm{d}x_0\mathrm{d}y_0\end{aligned} \tag{6.57}$$

Additionally, considering the estimated stochastic model of $\mathrm{loc}_c(t) = \big(x_c(t), y_c(t)\big)$ in Relation 6.55, we obtain the following relations:

$$\begin{aligned}
&\mathbf{Pr}\Big[G \in B\big(\mathrm{loc}_c(t), r\big)\Big| \begin{matrix}\mathrm{loc}_c(0) = (x_0, y_0)\\ \wedge \mathrm{loc}_c(t) \in S_c\end{matrix}\Big]\\
&= \mathbf{Pr}\Big[\mathrm{loc}_c(t) \in B\big(G, r\big)\Big| \begin{matrix}\mathrm{loc}_c(0) = (x_0, y_0)\\ \wedge \mathrm{loc}_c(t) \in S_c\end{matrix}\Big]\\
&= \iint_{\substack{(x,y)\\ \in B(G,r)}} \mathbf{Pr}\Big[\big(x_c(t), y_c(t)\big) = (x, y)\Big|\big(x_c(t), y_c(t)\big) \in S_c\Big]\\
&\leq \frac{\iint_{\substack{(x,y)\\ \in B(G,r)}} \mathbf{Pr}\Big[\big(x_c(t), y_c(t)\big) = (x, y)\Big]}{\mathbf{Pr}\Big[\big(x_c(t), y_c(t)\big) \in S_c\Big]}
\end{aligned}$$

Since for every $t \leq T$, $x_c(t)$ and $y_c(t)$ are normal distributed random variables of mean x_0 and y_0, respectively, this is the case that $(O = \mathrm{loc}_c(0) = (x_0, y_0))$:

$$\iint_{\substack{(x,y)\\ \in B(G,r)}} \mathbf{Pr}\Big[\big(x_c(t), y_c(t)\big) = (x, y)\Big] \leq \iint_{\substack{(x,y)\\ \in B(O,r)}} \mathbf{Pr}\Big[\big(x_c(t), y_c(t)\big) = (x, y)\Big]$$

which implies that

$$\begin{aligned}
\mathbf{Pr}\Big[G \in B\big(\mathrm{loc}_c(t), r\big)\Big| \begin{matrix}\mathrm{loc}_c(0) = (x_0, y_0)\\ \wedge \mathrm{loc}_c(t) \in S_c\end{matrix}\Big] &\leq \frac{\iint_{\substack{(x,y)\\ \in B(O,r)}} \mathbf{Pr}\Big[\big(x_c(t), y_c(t)\big) = (x, y)\Big]}{\mathbf{Pr}\Big[\big(x_c(t), y_c(t)\big) \in S_c\Big]} \qquad (6.58)\\
&\leq \frac{\int_{r'=0}^{r}\int_{\theta=0}^{2\pi} \frac{r'}{2\pi\sigma^2} \cdot \mathrm{e}^{-\frac{r'^2}{2\sigma^2 t}}\,\mathrm{d}r'\mathrm{d}\theta}{\mathbf{Pr}\Big[\big(x_c(t), y_c(t)\big) \in S_c\Big]}
\end{aligned}$$

Here, we make a claim which will be proved later:

$$B\big(O, r_{\min}\big) \subseteq S_c \qquad (6.59)$$

where

$$r_{\min} = \frac{\mu\gamma_n}{2\sin\left(\frac{2\pi}{n}\right)} \qquad (6.60)$$

Regarding Relation 6.59, we obtain the following inequality:

$$\mathbf{Pr}\left[G \in B\big(\mathrm{loc}_c(t), r\big)\middle| \begin{array}{c}\mathrm{loc}_c(0) = (x_0, y_0)\\ \wedge\mathrm{loc}_c(t) \in S_c\end{array}\right] \leq \frac{\int_{r'=0}^{r}\int_{\theta=0}^{2\pi}\frac{r'}{\sqrt{2\pi t\sigma^2}}\cdot \mathrm{e}^{-\frac{r'^2}{2t\sigma^2}}\mathrm{d}r'\mathrm{d}\theta}{\mathbf{Pr}\Big[\big(x_c(t), y_c(t)\big) \in B\big(O, r_{\min}\big)\Big]}$$
$$\leq \frac{\int_{r'=0}^{r}\int_{\theta=0}^{2\pi}\frac{r'}{\sqrt{2\pi t\sigma^2}}\cdot \mathrm{e}^{-\frac{r'^2}{2t\sigma^2}}\mathrm{d}r'\mathrm{d}\theta}{\int_{r'=0}^{r_{\min}}\int_{\theta=0}^{2\pi}\frac{r'}{\sqrt{2\pi t\sigma^2}}\cdot \mathrm{e}^{-\frac{r'^2}{2t\sigma^2}}\mathrm{d}r'\mathrm{d}\theta} \tag{6.61}$$
$$\leq \frac{1-e^{-\frac{r^2}{2t\sigma^2}}}{1-e^{-\frac{r_{\min}^2}{2t\sigma^2}}}$$

Using Eq. 6.57 and Inequality 6.61, we conclude that

$$\mathbf{Pr}\Big[G \in B\big(\mathrm{loc}_c(t), r\big)\Big|\mathrm{loc}_c(t) \in S_c\Big] \leq \frac{1-e^{-\frac{r^2}{2t\sigma^2}}}{1-e^{-\frac{r_{\min}^2}{2t\sigma^2}}} \times \iint_{\substack{(x_0,y_0)\\ \in S_c}} \mathbf{Pr}\big[\mathrm{loc}_c(0) = (x_0, y_0)\big]\mathrm{d}x_0\mathrm{d}y_0 \tag{6.62}$$

Regarding Lemma 6.6, we replace $\mathbf{Pr}\big[\mathrm{loc}_c(0) = (x_0, y_0)\big]$ by its upper bound in Inequality 6.62:

$$\mathbf{Pr}\Big[G \in B\big(\mathrm{loc}_c(t), r\big)\Big|\mathrm{loc}_c(t) \in S_c\Big] \leq \frac{1-e^{-\frac{r^2}{2t\sigma^2}}}{1-e^{-\frac{r_{\min}^2}{2t\sigma^2}}} \times \iint_{\substack{(x_0,y_0)\\ \in S_c}} \mathbf{Pr}\Big[\bigwedge_{i=0}^{m-1} X_i' = \mathrm{dist}((x_0, y_0), l_i')\Big]\mathrm{d}x_0\mathrm{d}y_0 \tag{6.63}$$

where line l_i' and function dist are defined as the same as what mentioned previously. In addition, similar to the proof of Theorem 6.1, we can get the following inequality:

$$\mathbf{Pr}\left[\bigwedge_{i=0}^{m-1} X_i' = \mathrm{dist}((x_0, y_0), l_i')\right] \leq \zeta_n\left(\frac{r}{\mu}\right)^3 \tag{6.64}$$

for some real positive sequence ζ_n. Inequalities 6.63 and 6.64 imply that

$$\mathbf{Pr}\Big[G \in B\big(\mathrm{loc}_c(t), r\big)\Big|\mathrm{loc}_c(t) \in S_c\Big] \leq \frac{1-e^{-\frac{r^2}{2t\sigma^2}}}{1-e^{-\frac{r_{\min}^2}{2t\sigma^2}}} \times \zeta_n\left(\frac{r}{\mu}\right)^3 |S_c|$$

It is easy to see that Lemma 6.8 is also true for the mobile case, i.e., area S_c belongs to ball $B\big(O, \mu \sin(\frac{\pi}{n})\big)$. As the result, this is the case that

$$
\begin{aligned}
|S_c| &\leq \left|B\left(O, \mu \sin\left(\frac{\pi}{n}\right)\right)\right| \\
&\leq \pi\mu^2 \sin^2\left(\frac{\pi}{n}\right)
\end{aligned}
$$

Henceforth, we obtain a lower bound for the instantaneous privacy level of mobile node c:

$$
\lambda(t) \leq \pi\mu^2 \sin^2\left(\frac{\pi}{n}\right) \zeta_n \frac{1 - e^{-\frac{r^2}{2t\sigma^2}}}{1 - e^{-\frac{r_{\text{min}}^2}{2t\sigma^2}}} \times \left(\frac{r}{\mu}\right)^3
$$

or,

$$
\lambda(t) \leq \zeta_n' \frac{1 - e^{-\frac{r^2}{2t\sigma^2}}}{1 - e^{-\frac{r_{\text{min}}^2}{2t\sigma^2}}} \times \left(\frac{r^3}{\mu}\right) \tag{6.65}
$$

for some positive real sequence ζ_n'.

To complete our analysis, we need to show Relation 6.59. Remember the notation X_i' which specifies the Euclidean distance between the static node's location O and the ith edge of the polygon S_c for every $i = 0, \ldots, m-1$. As mentioned before, for the mobile case, we change the second line of Algorithm 6.1 to Expression 6.56. The change we made in this algorithm will increase the minimum possible value of random variable X_i' from zero to r_{min}:

$$
r_{\text{min}} = \frac{\mu\gamma_n}{2\sin\left(\frac{2\pi}{n}\right)}
$$

Henceforth, we conclude Relation 6.59.

6.6.4 Concealing the Movement Path

In order to preserve the location privacy of a mobile node, not only we need to hide its instantaneous location, but we have to conceal its movement path to some extent. In our stochastic scheme, we quantify the privacy level of the node's path by calculating a probabilistic low threshold for random variable T that is the length of the time interval in which function MOBILEZONEUPDATER (Algorithm 6.2) keeps the anonymity zone unchanged.

As mentioned before, $B(O, r_{\text{min}}) \subseteq S_c$. This implies that

$$
\sup_{t \leq t'} \left\{ ||\text{loc}_c(t) - \text{loc}_c(0)|| \right\} \leq r_{\text{min}} \rightarrow T \geq t'
$$

Subsequently, we obtain the following proposition:

$$\sup_{t \le t'} \{x_c(t) - x_c(0)\} \le \frac{r_{\min}}{\sqrt{2}} \wedge \sup_{t \le t'} \{y_c(t) - y_c(0)\} \le \frac{r_{\min}}{\sqrt{2}} \tag{6.66}$$
$$\rightarrow T \ge t'$$

Now, we defined processes M_t and M'_t in the following form:

$$\begin{cases} M_{t'} = \sup_{t \le t'} \{x_c(t) - x_c(0)\} \\ M'_{t'} = \sup_{t \le t'} \{y_c(t) - y_c(0)\} \end{cases} \tag{6.67}$$

Concerning Proposition 6.66, we obtain the following inequality:

$$\mathbf{Pr}[T \ge t'] \ge \mathbf{Pr}\left[M_{t'} \le \frac{r_{\min}}{\sqrt{2}}\right] \times \mathbf{Pr}\left[M'_{t'} \le \frac{r_{\min}}{\sqrt{2}}\right] \tag{6.68}$$

Processes M_t and M'_t, respectively, represent the running maximum[8] of processes $\big(x_c(t) - x_c(0)\big)$ and $\big(y_c(t) - y_c(0)\big)$ which have been previously estimated by two Brownian motion processes of variance σ^2. As the result, this is the case that

$$\begin{cases} \mathbf{Pr}[M_t \le m] = \operatorname{erf}(\frac{m}{\sqrt{2t\sigma^2}}) \\ \mathbf{Pr}[M'_t \le m] = \operatorname{erf}(\frac{m}{\sqrt{2t\sigma^2}}) \end{cases} \tag{6.69}$$

Consequently, we obtain a probabilistic low threshold for random variable T:

$$\mathbf{Pr}[T \ge t'] \ge \operatorname{erf}^2\left(\frac{r_{\min}}{2\sqrt{t'\sigma^2}}\right) \tag{6.70}$$

6.7 Summary and Conclusion

In this chapter, we defined an initial anonymity zone for any static node first. Then, we shrank the zone using some geometric deductions. As mentioned, the shrunk anonymity zone is a convex polygon. We proposed an approach for finding the stochastic distribution of the node over the obtained anonymity zone. Additionally, we described the trade-off existing between the privacy level and the error tolerance of our scheme by obtaining thresholds for both the node's privacy level and the error

[8] Running maximum M_t of the Brownian motion process B_t is a random process which has the following cumulative density function at the arbitrary time $t > 0$ (σ^2 represents the variance of process B_t): $F_{M_t}(m) = \operatorname{erf}(\frac{m}{\sqrt{2t\sigma^2}})$ for every $m \ge 0$.

tolerance. Moreover, we extended our scheme for a mobile node with random walk on the 2D plane. In the mobile version, our scheme guarantees a specified minimum value for the location privacy level while assuring to hide the node's movement path.

References

1. R. Dewri, Location privacy and attacker knowledge: who are we fighting against? Lecture notes of the institute for computer sciences. Social Informatics and Telecommunications Engineering, vol. 96 (2012), pp. 96–115
2. M. Li, S. Salinas, A. Thapa, P. Li, n-CD: A Geometric Approach to Preserving Location Privacy in Location-Based Services, in *Proceedings of IEEE INFOCOM* (2013)
3. M. Gruteser, D. Grunwald, Anonymous usage of location-based services through spatial and temporal cloaking, in *ACM Mobisys'03*, May 2003
4. B. Gedik, L. Liu, Protecting location privacy with personalized kanonymity: architecture and algorithms. IEEE Trans. Mobile Comput. **7**(1), 118 (2008)
5. J. Meyerowitz, R.R. Choudhury, Hiding stars with fireworks: location privacy through camouflage, in *Proceedings of ACM MobiCom* (Beijing, China, Sept 2009)
6. M.F. Mokbel, C.Y. Chow, W.G. Aref, The new casper: query processing for location services without compromising privacy, in *Proceedings of VLDB*, 2006
7. P. Kalnis, G. Ghinita, K. Mouratidis, D. Papadia, Preventing location-based identity inference in anonymous spatial queries. IEEE Trans. Knowl. Data Eng. **19**(12), 1719–1733 (2007)
8. B. Gedik, L. Liu, Location privacy in mobile systems: a personalized anonymization model, in *Proceedings of IEEE ICDCS* (Columbus, Ohio, June 2005)
9. C.-Y. Chow, M.F. Mokbel, X. Liu, A peer-to-peer spatial cloaking algorithm for anonymous location-based service, in *Proceedings of ACM GIS* (Arlington, Virginia, Nov 2006)
10. A. Beresford, F. Stajano, Location privacy in pervasive computing. IEEE Pervasive Comput. **2**(1), 4655 (2003)
11. B. Hoh, M. Gruteser, H. Xiong, A. Alrabady, Preserving privacy in GPS traces via uncertainty-aware path cloaking, in *Proceedings of ACM CCS 2007* (Alexandria, US, Jan 2007)
12. H. Kido, Y. Yanagisawa, T. Satoh, An anonymous communication technique using dummies for location-based services, in *Proceedings of IEEE ICPS* (Santorini, Greece, July 2006)
13. H. Lu, C.S. Jensen, M.L. Yiu, Pad: privacy-area aware, dummy-based location privacy in mobile services, in *Proceedings of ACM MobiDE* (Vancouver, Canada, June 2008)
14. M. Duckham, L. Kulik, A formal model of obfuscation and negotiation for location privacy, in *Proceedings of international conference on pervasive computing* (Munich, Germany, May 2005)
15. C.A. Ardagna, M. Cremonini, S.D.C. di Vimercati, P. Samarati, An obfuscation-based approach for protecting location privacy. IEEE Trans. Dependable Secure Comput. **8**(1), 1327 (2011)
16. A. Pingley, W. Yu, N. Zhang, X. Fu, W. Zhao, Cap: a context-aware privacy protection system for location-based services, in *Proceedings of IEEE ICDCS* (Montreal, Canada, June 2009)
17. M. Damiani, E. Bertino, C. Silvestri, Probe: an obfuscation system for the protection of sensitive location information in lbs, Technical Report 2001–145, CERIAS, 2008

Erratum to: Expectation–Maximization for Acoustic Source Localization

Sitharama S. Iyengar, Kianoosh G. Boroojeni and N. Balakrishnan

Erratum to:
Chapter 3 in: S. S. Iyengar et al., *Mathematical Theories of Distributed Sensor Networks*, DOI 10.1007/978-1-4419-8420-3_3

The footnote on opening page of Chapter 3 should be read as below:

> "© 2010 IEEE. Reprinted, with permission, from "Robust Expectation–Maximization Algorithm for Multiple Wideband Acoustic Source Localization in the Presence of Nonuniform Noise Variances," Lu Lu, Hsiao-Chun Wu, Senior Member, IEEE, Kun Yan, and S. S. Iyengar, Fellow, IEEE, IEEE Sensors Journal, Vol. 11, No. 3, March 2011."

instead of

> "This chapter has been reprinted with permission from "Robust Expectation–Maximization Algorithm for Multiple Wideband Acoustic Source Localization in the Presence of Nonuniform Noise Variances,"Lu Lu, Hsiao-Chun Wu, Senior Member, IEEE, Kun Yan, and S. S. Iyengar, Fellow, IEEE, IEEE Sensors Journal, Vol. 11, No. 3, March 2011."

The online version of the original chapter can be found under DOI 10.1007/978-1-4419-8420-3_3

S. S. Iyengar (✉) · K. G. Boroojeni
Florida International University, Miami, FL, USA
e-mail: iyengar@cis.fiu.edu

K. G. Boroojeni
e-mail: kghol002@fiu.edu

N. Balakrishnan
Indian Institute of Science, Bangalore, India
e-mail: balki@serc.iisc.ernet.in

S. S. Iyengar et al., *Mathematical Theories of Distributed Sensor Networks*,
DOI: 10.1007/978-1-4419-8420-3_7, © Springer Science+Business Media New York 2014

Bibliography

1. S.S. Iyengar, R.L. Kashyap, R.N. Madan, Distributed sensor networks-introduction to the special section, IEEE Trans. Syst. Man Cybern. **21**(5), (1991)
2. R.R. Brooks, S.S. Iyengar, *Multi-sensor fusion: fundamentals and applications with software*. (Prentice Hall PTR, 1998)
3. R.F. Sprouil, D. Cohen, High level protocols in *Proceedings of IEEE, special issue on packet communication networks*, Nov 1978
4. D.B. Reid, An algorithm for tracking multiple targets, IEEE Trans. Autom. Contr. **AC-24** (1979)
5. S. Mori, C.Y. Chong, R.P. Wishner, E. Tse, Multi-target multi-sensor tracing problems: a general bayesian approach, in *Proceedings of American Control Conference*. (San Francisco, CA, 1983)
6. C.Y. Chong, S. Mori, Hierarchical multi-target tracking and classification—a Baysian approach, in *Proceedings of American Control Conference*. (San Diego, CA, 1984)
7. C.Y. Chong, E. Tse, S. Mori, Distributed estimation in networks, in *Proceedings of American Control Conference*. (San Francisco, CA, 1983)
8. C.Y. Chong, et al., Distributed hypothesis testing in distributed sensor networks, Artificial Intelligence & DS, Tech. Rep TP-1048-02, July 1984
9. J. O'Rourke, Art gallery theorems and algorithms. (Oxford University Press, 1987)
10. E.H. Callaway, *Wireless Sensor Networks: Architectures and Protocols*. (Auerbach Publications, New York, 2003)
11. H. Karl, A. Willig, *Protocols and Architectures for Wireless Sensor Networks*. (Wiley, New York, 2005)
12. I.F. Akyildiz, W. Su, Y. Sankarasubramaniam, E. Cayirci, Wireless sensor networks: a survey. Comput. Netw. **38**, 393–422 (2002)
13. A. Rahman, A. El Saddik, W. Gueaieb, Wireless Sensor Network Transport Layer: State of the Art, vol. 21, Sensors, *Lecture Notes in Electrical Engineering*. (Springer, Berlin, 2008), pp. 221–245
14. S. Kumar, T.H. Lai, and A. Arora, Barrier coverage with wireless sensor networks, in *Proceedings of the 11th Annual International Conference on Mobile Computing and Networking*. (Cologne, Germany, 2005), pp. 284–298
15. D.W. Gage, Command control for many-robot systems in *Proceedings of Nineteenth Annual AUVS Technical Symposium*, 1992, pp. 22–24
16. Hogeschool zeeland, Pipes, tubes, machinery and steam turbine at a power plant [Online] (2007), available http://goo.gl/fRp3c

S. S. Iyengar et al., *Mathematical Theories of Distributed Sensor Networks*,
DOI: 10.1007/978-1-4419-8420-3,

17. J.-H. Kim, G. Sharma, S. Iyengar, FAMPER: a fully autonomous mobile robot for pipeline exploration, in *Proceedings of IEEE International Conference on Industrial Technology* (2010), pp. 517–523
18. B. Ben-Moshe, M.J. Katz, J.S. Mitchell, A constant-factor approximation algorithm for optimal terrain guarding, in *Proceedings of ACM/SIAM Symposium on Discrete Algorithms* (2005), pp. 515–524
19. V. Chvatal, A combinatorial theorem in plane geometry. J. Comb. Theory **18**, 39–41 (1975)
20. J.O. Rourke, Galleries need fewer mobile guards: a variation on Chvatal's theorem. Geom. Dedicata **14**, 273–283 (1983)
21. J. Urrutia, Art gallery and illumination problems, *Handbook of Computational Geometry*, Amsterdam (North-Holland, The Netherlands, 2000)
22. J.O. Rourke, An alternate proof of the rectilinear art gallery theorem. J. Geom. **21**, 118–130 (1983)
23. J. O'Rourke, *Art Gallery Theorems and Algorithms* (Oxford University Press, London, UK, 1987)
24. I. Bjorling-Sachs, D. Souvaine, An efficient algorithm for guard placement in polygons with holes. Discr. Comput. Geom. **13**, 77–109 (1995)
25. E. Gyori, F. Hoffmann, K. Kriegel, T. Shermer, Generalized guarding and partitioning for rectilinear polygons. Comput. Geom. **6**(1), 21–44 (1996)
26. S. Fisk, Ashort proof of Chvatal's watchman theorem. J. Comb. Theor. Ser. B **24**(3), 374 (1978)
27. D. Avis, G. Toussaint, An efficient algorithm for decomposing a polygon into star-shaped polygons. Pattern Recognit. **13**, 395–398 (1981)
28. J. O'Rourke, K. Supowit, Some NP-hard polygon decomposition problems. IEEE Trans. Inf. Theory **29**, 181–190 (1983)
29. D.T. Lee, A.K. Lin, Computational complexity of art gallery problems. IEEE Trans. Inf. Theory **32**(2), 276–282 (1986)
30. D. Schuchardt, H.-D. Hecker, Two np-hard art-gallery problems for ortho-polygons. Math. Logic. Quart. **41**, 261–267 (1995)
31. M.J. Katz, G.S. Roisman, On guarding the vertices of rectilinear domains. Comput. Geom. Theory Appl. **39**(3), 219–228 (2008)
32. A. Efrat, S. Har-Peled, Guarding galleries and terrains. Inf. Process. Lett. **100**(6), 238–245 (2006)
33. J.-M. Lien, Approximate star-shaped decomposition of point set data, in *Eurographics Symposium on Point-Based Graphics*, Prague, Czech Republic (2007)
34. A. Nassiraei, Y. Kawamura, A. Ahrary, Y. Mikuriya, K. Ishii, Concept and design of a fully autonomous sewer pipe inspection mobile robot "KANTARO", in *Proceedings of the IEEE International Conference on Robotics and Automation*, 10–14 Apr 2007, pp. 136–143
35. J.Y. Choi, H. Lim, B.-J. Yi, Semi-automatic pipeline inspection robot systems, in *Proceedings of the SICE-ICASE International Joint Conference* (2006), pp. 2266–2269
36. S. Costello, D. Chapman, C. Rogers, N. Metje, Underground asset location and condition assessment technologies. Tunn. Undergr. Space Technol. **22**(5–6), 524–542 (2007)
37. J.-H. Kim, G. Sharma, N. Boudriga, S. Iyengar, Ramp system for proactive pipeline monitoring, in *Proceedings of the International Conference on Communication Systems, Network* (2010), pp. 1–2
38. J. Kim, G. Sharma, N. Boudriga, S. Iyengar, Spamms: asensor-based pipeline autonomous monitoring and maintenance system, in *Proceedings of the International Conference on Communication Systems, Network* (2010), pp. 1–10
39. R. Wirahadikusumah, D.M. Abraham, T. Iseley, R.K. Prasanth, Assessment technologies for sewer system rehabilitation. Autom. Construct. **7**(4), 259–270 (1998)
40. D.-H. Koo, S.T. Ariaratnam, Innovative method for assessment of underground sewer pipe condition. Autom. Construct. **15**(4), 479–488 (2006)
41. O. Duran, K. Althoefer, L. Seneviratne, State of the art in sensor technologies for sewer inspection. IEEE Sens. J. **2**(2), 73–81 (2002)

42. W.W. Zhang, B.H. Zhuang, Non-contact laser inspection for the inner wall surface of a pipe. Meas. Sci. Technol. **9**(9), 1380 (1998)
43. J. Moraleda, A. Ollero, M. Orte, A robotic system for internal inspection of water pipelines. IEEE Robot. Autom. Mag. **6**(3), 30–41 (1999)
44. M. Silk, The determination of crack penetration using ultrasonic surface waves. NDT Int. **9**(6), 290–297 (1976)
45. D. Levesque, M. Ochiai, A. Blouin, R. Talbot, A. Fukumoto, J.-P. Monchalin, Laser-ultrasonic inspection of surface-breaking tight cracks in metals using SAFT processing, in *Proceedings of the IEEE Ultrasonics Symposium*, vol. 1, 8–11 Oct 2002, pp. 753–756
46. J. Hertzberg, F. Kirchner, Landmark-based autonomous navigation in sewerage pipes, in *Proceedings of the 1st Euromicro Workshop on Advanced Mobile Robots*, Oct 1996, pp. 68–73
47. H. Streich, O. Adria, Software approach for the autonomous inspection robot MAKRO. Proc. IEEE Int. Conf. Robot. Autom. **4**, 3411–3416 (2004)
48. J.M. Mirats Tur, W. Garthwaite, Robotic devices for water main in-pipe inspection: a survey. J. Field Robot. **27**, 491–508 (2010)
49. T. He, L. Hong, D. Chen, Z. Liang, Reliable path for virtual endoscopy: ensuring complete examination of human organs. IEEE Trans. Vis. Comput. Graphics **7**(4), 333–342 (2001)
50. D.-G. Kang, J.B. Ra, A new path planning algorithm for maximizing visibility in computed tomography colonography. IEEE Trans. Med. Imag. **24**(8), 957–968 (2005)
51. N. Cornea, D. Silver, P. Min, Curve-skeleton properties, applications, and algorithms. IEEE Trans. Vis. Comput. Graphics **13**(3), 530–548 (2007)
52. T.K. Dey, J. Sun, Defining and computing curve-skeletons with medial geodesic function, in *Proceedings of the 4th Eurographics Symposium on Geometry Processing* (Eurographics Association Aire-la-Ville, Switzerland, 2006), pp. 143–152
53. J. Kahn, M. Klawe, D. Kleitman, Traditional galleries require fewer watchmen. SIAM J. Algebraic Discr. Methods **4**(2), 194–206 (1983)
54. D.S. Johnson, Approximation algorithms for combinatorial problems, in *Proceedings of the 5th Annual ACM Symposium on Theory Computer*, New York (1973), pp. 38–49
55. TOMLAB v3.0 User's Guide (Department of Mathematics and Physics, Malardalen University, Vasteras, 2001), Technical report IMa-TOM-2001-01
56. H. Hoppe, Progressive meshes, in *Proceedings of the SIGGRAPH*, New York (1996), pp. 99–108
57. X. Li, Z. Yin, L. Wei, S. Wan, W. Yu, M. Li, Symmetry and template guided completion of damaged skulls. Comput. Graph. **35**, 885–893 (2011)
58. R. Finkel, J. Bentley, Quad trees: a data structure for retrieval on composite keys. Acta Inf. **4**, 1–9 (1974)
59. T. Kuroki, K. Terabayashi, K. Umeda, Construction of a compact range image sensor using multi-slit laser projector and obstacle detection of a humanoid with the sensor, in *Proceedings of the IEEE/RSJ International Conference on Intelligence Robots Systems*, Oct 2010, pp. 5972–5977
60. J. Thielemann, G. Breivik, A. Berge, Pipeline landmark detection for autonomous robot navigation using time-of-flight imagery, in *Proceedings of the IEEE Computer Society Conference on Computer Vision and Pattern Recognition, Workshops*, 23–28 Jun 2008, pp. 1–7
61. K. Yan, H.-C. Wu, S.S. Iyengar, Robustness analysis of source localization using Gaussianity measure, in *Proceedings of the IEEE Global Telecommunication Conference*, Nov 2008, pp. 1–5
62. H. Krim, M. Viberg, Two decades of array signal processing research: the parametric approach. IEEE Signal Process. Mag. **13**(4), 67–94 (1996)
63. J.C. Chen, R.E. Hudson, K. Yao, Maximum-likelihood source localization and unknown sensor location estimation for wideband signals in the near-field. IEEE Trans. Signal Process. **50**(8), 1843–1854 (2002)
64. J.C. Chen, R.E. Hudson, K. Yao, Source localization of a wideband source using a randomly distributed beam forming sensor array, in *Proceedings of the International Society, Information Fusion* (2001), pp. TuC1: 11–18

65. P. Stoica, A. Nehorai, Music, maximum likelihood and Cramer-Rao bound. IEEE Trans Acoust. Speech Signal Process. **37**(5), 720–741 (1989)
66. P. Stoica, A. Nehorai, Performance study of conditional and unconditional direction-of-arrival estimation. IEEE Trans Acoust. Speech Signal Process. **38**(10), 1783–1795 (1990)
67. C.E. Chen, F. Lorenzelli, R.E. Hudson, K. Yao, Maximum likelihood DOA estimation of multiple wideband sources in the presence of nonuniform sensor noise, EURASIP J. Adv. Signal Process. **2008** (2008)
68. M. Pesavento, A.B. Gershman, Maximum-likelihood direction- of-arrival estimation in the presence of unknown nonuniform noise. IEEE Trans. Signal Process. **49**(7), 1310–1324 (2001)
69. A.P. Dempster, N.M. Laird, D.B. Rubin, Maximum likelihood from incomplete data via the EM algorithm. J. Roy. Statist. Soc. **39**(1), 1–38 (1977)
70. M. Feder, E. Weinstein, Multipath and multiple source array processing via the EM algorithm, in *Proceedings of IEEE International Conference of Acoustics, Speech, Signal Processing*, vol. 11 (1986), pp. 2503–2506
71. M. Feder, E. Weinstein, Parameter estimation of superimposed signals using the EM algorithm. IEEE Trans. Acoust. Speech Signal Process. **36**(4), 477–489 (1988)
72. E. Weinstein, V. Oppenheim, M. Feder, Iterative and sequential algorithms for multisensor signal enhancement. IEEE Trans. Signal Process. **42**(4), 846–859 (1994)
73. P.J. Chung, J.F. Bohme, Recursive EM and SAGE algorithms, in *Proceedings of IEEE Workshop Statistics Signal Processing* (2001), pp. 540–543
74. P.J. Chung, J.F. Bohme, A.O. Hero, Tracking of multiple moving sources using recursive EM algorithm. EURASIP J. Appl. Signal Process. **1**(1), 50–60 (2005)
75. L. Frenkel, M. Feder, Recursive expectation-maximization (EM) algorithms for time-varying parameters with applications to multiple target tracking. IEEE Trans. Signal Process. **47**(2), 306–320 (1999)
76. M.I. Miller, D.R. Fuhrmann, Maximum-likelihood narrow-band direction finding and the EM algorithm. IEEE Trans. Acoust. Speech Signal Process. **38**(9), 1560–1577 (1990)
77. K.K. Mada, H.-C. Wu, EM algorithm for multiple wideband source, in *Proceedings of IEEE Global Telecommunication Conference* (2006), pp. 1–5
78. K.K. Mada, H.-C. Wu, S.S. Iyengar, Efficient and robust EM algorithm for multiple wideband source localization. IEEE Trans. Veh. Technol. **58**(6), 3071–3075 (2009)
79. L. Lu, H.-C. Wu, K. Yan, S.S. Iyengar, Robust expectation-maximization algorithm for multiple wideband acoustic source localization in the presence of nonuniform noise variances. IEEE Sens. J. **11**(3) (2011)
80. M. Allili, K. Mischaikow, A. Tannenbaum, Cubical homology and the topological classification of 2D and 3D imagery, in IEEE International Conference on Image Processing (2001), pp. 173–176
81. A. Ames, A homology theory for hybrid systems: hybrid homology. Lecture notes in computer science, vol. 3414 (2005), p. 86102
82. M. Batalin, G. Sukhatme, Spreading out: a local approach to multi-robot coverage, in *Proceedings of 6th International Symposium on Distributed Autonomous Robotic Systems*, Fukuoka, Japan (2002)
83. M. Batalin, M. Hattig, G. Sukhatme, Mobile robot navigation using a sensor network, in Proceedings of ICRA (2004)
84. K. Bekris, L. Kavraki, A review of recent results in robotic sensor networks. ACM Comput. Surv. **V(N)** (2005)
85. N. Bulusu, J. Heidelmann, D. Estrin, Adaptive beacon placement, in *Proceedings of Conference on Distributed Computing Systems* (2001)
86. P. Corke, R. Peterson, D. Rus, Localization and navigation assisted by cooperating networked sensors and robots. Int. J. Rob. Res. **24**(9), 771 (2005)
87. J. Cortes, S. Martinez, T. Karatas, F. Bullo, Coverage control for mobile sensing networks. IEEE Trans. Rob. Autom. **20**(2), 243–255 (2004)
88. C. Delfinado, H. Edelsbrunner, An incremental algorithms for Betti numbers of simplicial complexes on the 3-spheres. Comp. Aided Geom. Des. **12**(7), 771–784 (1995)

89. V. de Silva, G. Carlsson, Topological estimation using witness complexes, in Symposium on Point-Based Graphics, ETH Zurich (2004)
90. V. de Silva, R. Ghrist, Coverage in sensor networks via persistent homology. Alg. Geom. Topology. (to appear)
91. V. de Silva, R. Ghrist, A. Muhammad, Blind swarms for coverage in 2-d, in *Proceedings of Robotics: Systems and Science* (2005)
92. J. Eckhoff, Helly, Radon, and Caratheodory type theorems, Chap. 2.1, in *Handbook of Convex Geometry*, ed. by P.M. Gruber, J.M. Wills (North-Holland, Amsterdam, Netherlands, 1993), pp. 389–448
93. D. Estrin, D. Culler, K. Pister, G. Sukhatme, Connecting the physical world with pervasive networks. IEEE Pervasive Comput. **1**(1), 5969 (2002)
94. S. Fekete, A. Kröller, D. Pfisterer, S. Fischer, Deterministic boundary recognition and topology extraction for large sensor networks, in *Algorithmic Aspects of Large and Complex Networks* (2006)
95. D.W. Gage, Command control for many-robot systems. Nineteenth Annual AUVS Technical Symposium, Huntsville, Alabama, USA (1992), pp. 22–24
96. A. Galstyan, B. Krishnamachari, K. Lerman, S. Pattem, Distributed online localization in sensor networks using a moving target (preprint, 2003)
97. R. Ghrist, D. Lipsky, S. Poduri, G. Sukhatme, Node isolation in coordinate-free networks, in *Proceedings of the Sixth International Workshop on Algorithmic Foundations of Robotics* (2006)
98. M. Gromov, Hyperbolic groups, in *Essays in Group Theory* (MSRI Publication 8, Springer, 1987)
99. A. Hatcher, *Algebraic Topology* (Cambridge University Press, Cambridge, 2002)
100. C.-F. Hsin, M. Liu, Network coverage using low duty-cycled sensors: random and coordinated sleep algorithms, in *Proceedings of IPSN* (2004)
101. C.-F. Huang, Y.-C. Tseng, The coverage problem in a wireless sensor network, in ACM International Workshop on Wireless Sensor Networks and Applications (2003), pp. 115–121
102. T. Kaczynski, K. Mischaikow, M. Mrozek, *Computational Homology, Applied Mathematical Sciences*, vol. 157 (Springer, 2004)
103. H. Koskinen, On the coverage of a random sensor network in a bounded domain, in *Proceedings of 16th ITC Specialist, Seminar* (2004), pp. 11–18
104. S. Kumar, T.H. Lai, J. Balogh, On k-coverage in a mostly sleeping sensor network, in *Proceedings of 10th International Conference on Mobile Computing and Networking* (2004)
105. X.-Y. Li, P.-J. Wan, O. Frieder, Coverage in wireless ad-hoc sensor networks. IEEE Trans. Comput. **52**(6), 753–763 (2003)
106. B. Liu, D. Towsley, A study of the coverage of large-scale sensor networks, in IEEE International Conference on Mobile Ad-hoc and Sensor Systems (2004)
107. S. Meguerdichian, F. Koushanfar, M. Potkonjak, M. Srivastava, Coverage problems in wireless ad-hoc sensor network. IEEE INFOCOM (2001), pp. 1380–1387
108. K. Mischaikow, M. Mrozek, J. Reiss, A. Szymczak, Construction of symbolic dynamics from experimental time series. Phys. Rev. Lett. **82**(6), 1144 (1999)
109. R. Moses, D. Krishnamurthy, R. Patterson, A self-localization methods for wireless sensor networks. EURASIP J. Appl. Signal Proc. (2002)
110. A. Muhammad, R. Ghrist, M. Egerstedt, Beyond graphs: higher dimensional structures for networked control systems. Control Sys. Mag. (2006) (to appear)
111. A. Rao, S. Ratnasamy, C. Papdimitriou, S. Shenker, I. Stoika, Geographic routing without location information, in *Proceedings of ACM MOBICCOM* (2003)
112. S. Simic, S. Sastry, Distributed environmentalmonitoring using random sensor networks. Lecture notes in computer science, vol. 2634 (2003), pp. 582–592
113. L. Vietoris, Uber den höheren Zusammenhang kompakter Raume und eine Klasse von zusammenhangstreuen Abbildungen. Math. Ann. **97**, 454–472 (1927)
114. F. Xue, P.R. Kumar, The number of neighbors needed for connectivity of wireless networks. Wirel. Netw. **10**(2), 169–181 (2004)

115. H. Zhang, J. Hou, Maintaining coverage and connectivity in large sensor networks. International Workshop on Theoretical and Algorithmic Aspects of Sensor (Ad hoc Wireless and Peer-to- Peer Networks, Florida, 2004)
116. A. Zomorodian, G. Carlsson, Computing persistent homology, in *Proceeding of 20th ACM Symposium on Computational Geometry* (2004), pp. 346–356
117. Computational Geometry Algorithms Library, http://www.cgal/org/
118. Plex, version 2.1, Jan (2006), http://math.stanford.edu/comptop/programs/plex/
119. C. Gui, P. Mohapatra, Power conservation and quality of surveillance in target tracking sensor networks. In *Proceedings of the 10th Annual International Conference on Mobile Computing and Networking (Mobicom 04)*, Philadelphia, PA, USA, 26 Sept–01 Oct 2004, pp. 129–143
120. B. Liu, D. Towsley, A study of the coverage of large-scale sensor networks. In *Proceedings of 2004 IEEE International Conference on Mobile Ad-hoc and Sensor Systems*, Fort Lauderdale, FL, USA, 25–27 Oct 2004, pp. 475–483
121. Y. Zou, K. Chakrabarty, Sensor deployment and target localization in distributed sensor networks. ACM Trans. Embed. Comput. Syst. **3**, 61–91 (2004)
122. S.P. Fekete, A. Kroller, D. Pfisterer, S. Fischer, C. Buschmann, Neighborhood-based topology recognition in sensor networks, in *Proceedings of ALGOSENSORS*, Turku, Finland (2004), pp. 123–136
123. S.-P. Kuo, Y.-C. Tseng, F.-J. Wu, C.-Y. Lin, A probabilistic signal-strength-based evaluation methodology for sensor network deployment. In *Proceedings of 19th International Conference on Advanced Information Networking and Applications, AINA 2005*, Taipei, Taiwan, vol. 1, 28–30 March 2005, pp. 319–324
124. S. Ren, Q. Li, H. Wang, X. Chen, X. Zhang, A study on object tracking quality under probabilistic coverage in sensor networks. SIGMOBILE Mob. Comput. Commun. Rev. **9**, 73–76 (2005)
125. A. Muhammad, A. Jadbabaie, Dynamic coverage verification in mobile sensor networks via switched higher order Laplacians, in *Proceedings of Robotics: Science and Systems*, Atlanta, GA, USA (2007)
126. A. Tahbaz-Salehi, A. Jadbabaie, Distributed coverage verification in sensor networks without location information, in *Proceedings of 47th IEEE Conference on Decision and Control, CDC 2008*, Cancun, Mexico, 9–11 Dec 2008, pp. 4170–4176
127. P. Corke, R. Peterson, D. Rus, Localization and navigation assisted by networked cooperating sensors and robots. Int. J. Rob. Res. **24**, 771–786 (2005)
128. H. Zhang, J. Hou, Maintaining sensing coverage and connectivity in large sensor networks. Wirel. Ad Hoc Sens. Netw. **1**, 89–123 (2005)
129. C.-F. Hsin, M. Liu, Network coverage using low duty-cycled sensors: random and coordinated sleep algorithms, in *Proceedings of the 3rd International Symposium on Information Processing in Sensor Networks, IPSN'04*, New York, NY, USA (2004), pp. 433–442
130. M. Abdelkader, M. Hamdi, N. Boudriga, Using higher-order Voronoi tessellations for WSN-based target tracking, in *Proceedings of the IASTED International Symposium, Distributed Sensor Networks, DSN'08*, Calgary, AB, Canada, 21–24 Sept 2008, pp. 430–435
131. S. Meguerdichian, F. Koushanfar, M. Potkonjak, M. Srivastava, Coverage problems in wireless ad-hoc sensor networks, in *Proceedings of INFOCOM 2001 Twentieth Annual Joint Conference of the IEEE Computer and Communications Societies*, Anchorage, AK, USA, vol. 3, 22–26 April 2001, pp. 1380–1387
132. X.-Y. Li, P.-J. Wan, O. Frieder, Coverage in wireless ad hoc sensor networks. Comput. IEEE Trans. **52**, 753–763 (2003)
133. C.-F. Huang, Y.-C. Tseng, The coverage problem in a wireless sensor network, in *Proceedings of The 2nd ACM International Conference on Wireless Sensor Networks and Applications, WSNA'03*, San Diego, CA, USA, 19 Sept 2003, pp. 115–121
134. H. Koskinen, On the coverage of a random sensor network in a bounded domain, in *Proceedings of the 16th ITC Specialist Seminar on Performance Evaluation of Wireless and Mobile Systems*, Antwerp, Belgium, 31 Aug–02 Sept 2004, pp. 11–18

135. R. Ghrist, A. Muhammad, Coverage and hole-detection in sensor networks via homology, in *Proceedings of the 4th International Symposium on Information Processing in Sensor Networks, IPSN'05*, Los Angeles, CA, USA, 25–27 April 2005, pp. 254–260
136. A. Gramain, *Topology of Surfaces* (BCS Associates, Moscow, ID, USA, 1984)
137. A. Hatcher, *Algebraic Topology* (Cambridge University Press, Cambridge, UK, 2001)
138. V. De Silva, R. Ghrist, Coverage in sensor networks via persistent homology. Alg. Geom. Topology **7**, 339–358 (2007)
139. V. De Silva, R. Ghrist, Coordinate-free coverage in sensor networks with controlled boundaries via homology. Int. J. Rob. Res. **25**, 1205–1222 (2006)
140. R. Dewri, Location privacy and attacker knowledge: who are we fighting against?. Lecture notes of the institute for computer sciences. Social Informatics and Telecommunications Engineering **96**, 96–115 (2012)
141. M. Li, S. Salinas, A. Thapa, P. Li, n-CD: a geometric approach to preserving location privacy in location-based services, in *Proceedings of IEEE INFOCOM* (2013)
142. M. Gruteser, D. Grunwald, Anonymous usage of location-based services through spatial and temporal cloaking. ACM Mobisys'03 (2003)
143. B. Gedik, L. Liu, Protecting location privacy with personalized kanonymity: architecture and algorithms. IEEE Trans. Mob. Comput. **7**(1), 118 (2008)
144. J. Meyerowitz, R.R. Choudhury, Hiding stars with fireworks: location privacy through camouflage, in *Proceedings of ACM MobiCom*, Beijing, China (2009)
145. M.F. Mokbel, C.Y. Chow, W.G. Aref, The new casper: query processing for location services without compromising privacy, in *Proceedings of VLDB* (2006)
146. P. Kalnis, G. Ghinita, K. Mouratidis, D. Papadias, Preventing location-based identity inference in anonymous spatial queries. IEEE Trans. Knowl. Data Eng. **19**(12), 1719–1733 (2007)
147. B. Gedik, L. Liu, Location privacy in mobile systems: a personalized anonymization model, in *Proceedings of IEEE ICDCS*, Columbus, Ohio (2005)
148. C.-Y. Chow, M.F. Mokbel, X. Liu, A peer-to-peer spatial cloaking algorithm for anonymous location-based service, in *Proceedings of ACM GIS*, Arlington, Virginia (2006)
149. A. Beresford, F. Stajano, Location privacy in pervasive computing. IEEE Pervasive Comput. **2**(1), 4655 (2003)
150. B. Hoh, M. Gruteser, H. Xiong, A. Alrabady, Preserving privacy in gps traces via uncertainty-aware path cloaking, in *Proceedings of ACM CCS 2007*, Alexandria, VA, US (2007)
151. H. Kido, Y. Yanagisawa, T. Satoh, An anonymous communication technique using dummies for location-based services, in *Proceedings of IEEE ICPS*, Santorini, Greece (2006)
152. H. Lu, C.S. Jensen, M.L. Yiu, Pad: privacy-area aware, dummy-based location privacy in mobile services, in *Proceedings of ACM MobiDE*, Vancouver, Canada (2008)
153. M. Duckham, L. Kulik, A formal model of obfuscation and negotiation for location privacy, in *Proceedings of International Conference on Pervasive Computing*, Munich, Germany (2005)
154. C.A. Ardagna, M. Cremonini, S.D.C. di Vimercati, P. Samarati, An obfuscation-based approach for protecting location privacy. IEEE Trans. Dependable Secure Comput. **8**(1), 1327 (2011)
155. A. Pingley, W. Yu, N. Zhang, X. Fu, W. Zhao, Cap: a context-aware privacy protection system for location-based services, in *Proceedings of IEEE ICDCS*, Montreal, Canada (2009)
156. M. Damiani, E. Bertino, C. Silvestri, Probe: an obfuscation system for the protection of sensitive location information in lbs. Technical report 2001–145, CERIAS (2008)

MIX
Papier aus verantwortungsvollen Quellen
Paper from responsible sources
FSC® C105338

If you have any concerns about our products,
you can contact us on
ProductSafety@springernature.com

In case Publisher is established outside the EU,
the EU authorized representative is:
Springer Nature Customer Service Center GmbH
Europaplatz 3, 69115 Heidelberg, Germany

Printed by Libri Plureos GmbH
in Hamburg, Germany